Exercises in SEDIMENTOLOGY

Gerald M. Friedman
Rensselaer Polytechnic Institute

Kenneth G. Johnson
Skidmore College

1807 175 YEARS OF PUBLISHING 1982

JOHN WILEY & SONS
New York Chichester Brisbane Toronto Singapore

ISBN 0-471-87453-1
Printed in the United States of America

10 9 8 7 6 5 4 3 2 1

PREFACE

This manual contains sedimentology exercises for students; it is not a sedimentology reference manual. In the following pages you will find ten exercises, some of which require more than one laboratory session to complete.

The study of sedimentology involves consideration of sedimentary materials, sedimentary processes, and the various products of these processes, with the ultimate end in view of interpreting the three-dimensional relationships of sedimentary strata. This manual contains exercises designed to introduce the student to the procedures used in sedimentological analysis.

Exercises One through Eight deal with the compositional and textural nature of sedimentary particles and the myriad combinations in which such particles may accumulate in nature, whether as modern sedimentary deposits or their ancient equivalents - sedimentary rocks. Exercise Nine deals with the analysis of three-dimensional arrangements in which sedimentary rocks are found, and Exercise Ten is designed to simulate the geological and economic conditions encountered in exploration for oil and gas. In general, Exercises One through Ten proceed progressively from very specific laboratory collection and analysis of data to more general, office treatment and interpretation of field and exploration information.

An instructor's supplement has been prepared to accompany Exercises in Sedimentology in order to assist instructors in fitting the exercises into particular course contexts. All of the exercises in this manual were generated by the authors, with the exception of Exercise Ten. This exercise is a modification of a "petroleum game" devised by Shepard W. Lowman, late Professor of Geology at Rensselaer Polytechnic Institute.

Numerous references are listed within individual exercises and a short list of other references is included following Exercise Ten. The last section of the manual is a glossary of sedimentological terms. All of these sections combine to make the manual a valuable information source for the student after completion of formal in-course study of sedimentology.

The authors wish to acknowledge the contributions of John E. Sanders of Barnard College, who wrote parts of the material taken directly from Principles of Sedimentology by Friedman and Sanders and used for this manual; Yitzhak Levy of Weizmann Institute of Science and the Geological Survey of Israel, who wrote early versions of several of the exercises while a visiting professor at Rensselaer Polytechnic Institute; and Richard

H. Lindemann of Skidmore College, who critically read Exercises One through Four. Betty B. Platt of Skidmore College meticulously typed several manuscript drafts. Gene Tobler and Mary Curl of Rensselaer Polytechnic Institute provided photographic assistance. An NSF Undergraduate Development Grant supported early formulation by Friedman of several of the manual exercises.

Although this manual has been written to complement the text Principles of Sedimentology by Friedman and Sanders, it is designed to serve equally well for laboratory instruction in a course for which a text other than Principles of Sedimentology is being used.

Gerald M. Friedman
Kenneth G. Johnson

TABLE OF CONTENTS

	Page
LIST OF FIGURES	x
LIST OF TABLES	xii
LIST OF REPORT FORMS	xiii
EXERCISE 1 - SEDIMENTARY PARTICLES	
Introduction	2
Kinds of Sedimentary Particles	2
Group-One Particles: solid breakdown products of older deposits	2
Group-Two Particles: materials that are not solid breakdown products of older deposits	2
References Dealing with Sedimentary Particles	3
Equipment and Materials	4
Procedures	4
Procedure I: Staining Feldspar in Sediments	5
References Dealing with Staining of Feldspars	6
Procedure II: Examination of Fragments of Igneous, Metamorphic and Sedimentary Rock	6
Procedure III: Examination of Mud, Sand and Gravel	11
Procedure IV: Staining Calcite and Dolomite in Sediments	14
References Dealing with Staining of Carbonates	17
Procedure V: Examination of Samples Containing Carbonate Particles	17
References Dealing with the Identification of Carbonate-Sedimentary Particles	17
Other Examples of Group-Two Particles	18
EXERCISE 2 - TEXTURAL ANALYSIS OF SAND	
Introduction	24
Particle-Size Analysis	24
References Dealing with Sediment Particle-Size Analysis	26
Equipment and Materials	26
Determination of Particle-Size Distribution	27
Procedure I: Splitting the Sample	27
Procedure II: Sieving the Sample	28
Relationship of Particle Size to Mineralogical Composition	32
Procedure III: Study of Mineralogical Composition of Individual Size Fractions	

	Page
Study of Particle Shapes and Surface Textures	32
Procedure IV: Estimation of Sphericity	36
References Dealing with Particle Sphericity	36
Procedure V: Estimation of Roundness	38
References Dealing with Particle Roundness	38
Procedure VI: Determination of Surface Texture	39
References Dealing with Surface Texture	39
Procedure VII: Making a Particle-Size Standard	40
Synthesis of Data	40
EXERCISE 3 - SIZE ANALYSIS OF SILT AND CLAY	
Introduction	44
References Dealing with Size Analysis of Silt and Clay	44
Settling Velocity	45
Equipment and Materials	45
Procedure	46
The Pipette Method	48
Procedure I: Pipette Analysis	48
EXERCISE 4 - GRAPHIC PRESENTATION OF PARTICLE-SIZE DISTRIBUTION	
Introduction	52
References Dealing with Graphic Presentation of Particle-Size Distribution	52
Material	54
The Normal Distribution	54
The Histogram	55
Procedure I: Plotting Data as a Histogram	56
Variations in Shape of the Frequency Curve	56
Procedure II: Plotting Data as a Frequency-Distribution Curve	58
The Cumulative-Frequency Curve	62
Procedure III: Plotting Data as a Cumulative-Frequency Curve (Arithmetic Ordinate)	62
Procedure IV: Plotting Data as a Cumulative-Frequency Curve (Probability Ordinate)	66
Graphic Measures of the Frequency Distribution	68
Procedure V: Computation of Approximate Statistical Parameters of the Particle-Size Distribution	69
Interpretation of Data	69

	Page
Procedure VI: Interpretation of Particle-Size Distribution Data	75
Moment Measures	76
References Dealing with Moment Measures	82
EXERCISE 5 - HEAVY-MINERAL ANALYSIS AND PROVENANCE	
Introduction	86
Heavy Minerals	86
Caution	88
References Dealing with Heavy-Mineral Analysis	89
Equipment and Materials	90
Separation of Heavy Minerals	90
Procedure I: Separation of Heavy-Mineral Particles by the Gravity Method	91
Procedure II: Separation of Heavy-Mineral Particles by the Magnetic-Susceptibility Method	94
Determination of Provenance	95
Procedure III: Determination of Heavy-Mineral Composition	96
Procedure IV: Interpretation of Provenance	106
EXERCISE 6 - CARBONATE STAINING AND PEELS	
Introduction	108
References Dealing with Carbonate Staining and Peels	108
Equipment and Materials	109
Procedures	110
Procedure I: Slab Preparation	110
Procedure II: Etching the Slab	111
Procedure III: Staining the Slab	114
Procedure IV: Preparation of the Peel	114
EXERCISE 7 - DESCRIPTION AND CLASSIFICATION OF SEDIMENTARY ROCKS IN HAND SPECIMEN AND THIN SECTION: INTRABASINAL ROCKS	
Introduction	118
Limestones	119
Leighton-Pendexter Classification	120
Dunham Classification	127
Folk Classification	129
Problems and Limitations in Classifying Limestones	130
Techniques	131

	Page
Transfer of Lime Mud	131
Precipitation of Cryptocrystalline Cement	131
Recrystallization	132
References Dealing with Limestones	137
Dolostones	138
References Dealing with Dolostones	142
Authigenic Rocks	142
Chert	143
Evaporites	145
Phosphate Rock	147
Sedimentary-Iron Ores	148
Carbonaceous Rocks	150
References Dealing with Authigenic Rocks	152
Procedure	152

EXERCISE 8 - DESCRIPTION AND CLASSIFICATION OF SEDIMENTARY ROCKS IN HAND SPECIMEN AND THIN SECTION: EXTRABASINAL AND PYROCLASTIC ROCKS

Introduction	156
Shales	157
References Dealing with Shales	158
Sandstones	158
References Dealing with Sandstones	163
Conglomerates and Sedimentary Breccias	164
References Dealing with Conglomerates and Sedimentary Breccias	167
Pyroclastic Rocks	167
References Dealing with Pyroclastic Rocks	169
Procedure	170

EXERCISE 9 - FACIES ANALYSIS

Introduction	174
Depositional Environments and Facies	174
Materials	181
Procedures	181
Procedure I: Plotting Stratigraphic Cross-Section	189
Procedure II: Designation of Facies	190
Procedure III: Interpretation of Facies	190

	Page
EXERCISE 10 - OIL AND GAS EXPLORATION PROJECT	
Procedure	194
REFERENCES CITED	197
GLOSSARY	199

LIST OF FIGURES

Figure		Page
1-1	Photomicrographs of Particles which are Solid Breakdown Products of Older Deposits	7
1-2	Photomicrographs of Particles which are Solid Breakdown Products of Older Deposits	8
1-3	Photomicrographs of Particles which are Solid Breakdown Products of Older Deposits	12
1-4	Photomicrographs of Particles which are Solid Breakdown Products of Older Deposits	13
1-5	Photographs of Carbonate Particles that Grew as Solids in the Depositional Basin	15
1-6	Photomicrograph of Carbonate Particles that Grew as Solids in the Depositional Basin	16
1-7	Spherical and Ovoidal Pellets of Glauconite	19
1-8	Photomicrograph of Glass Shards of Tephra	20
1-9	Photomicrograph of Volcanic Ash	21
2-1	Mechanical Sample Splitter	27
2-2	Sieving Devices	29
2-3	Electronic Balance	31
2-4	Outlines of Six Roundness Classes	38
4-1	Normal Frequency-Distribution Curve	54
4-2	Histogram of Particle-Size Data Shown in Figure 4-1	56
4-3	Normal Frequency-Distribution Curves having Various Sorting Characteristics	58
4-4	Frequency-Distribution Curves having Various Skewnesses	59
4-5	Frequency Distributions of Sands from Continental Shelf off Long Island, New York, showing Various Modes	60
4-6	Cumulative-Frequency Curve having an Arithmetic Ordinate Scale	62
4-7	Cumulative-Frequency Curve having a Probability Ordinate Scale	64
4-8	Cumulative-Frequency Curve of Beach Sand, Drawn on Probability Paper	65
4-9	Cumulative-Frequency Curves, River Sand and Beach Sand	72
4-10	Scatter Plot, Skewness vs. Sorting - Beach and River Sands	73
4-11	Plot of Inclusive Graphic Standard Deviation and Inclusive Graphic Skewness - Beach- and Inland-Dune Sands	74
4-12	Plot of Inclusive Graphic Standard Deviation and Inclusive Graphic Skewness - Inland- and Nearshore Dune Sands	76
4-13	Moments in Mechanics and Statistics	77
4-14	Moment about Mean	79
4-15	Frequency-Distribution Curves having Various Kurtoses	82

Figure		Page
4-16	Scatter Plot, Standard Deviation versus Third Moment (mean cubed deviation), for Beach Sands (of kind specified in text) and River Sands	83
5-1	Opaque Heavy Minerals, Including Rutile and Leucoxene	88
5-2	Simple Apparatus for Heavy-Mineral Separation by Gravity Method	91
5-3	Frantz Isodynamic Magnetic Separator	93
6-1	Recommended Staining Procedure	112
7-1	Polished Slab of Carbonate Conglomerate	122
7-2	Spines of Echinoid	122
7-3	Polished and Etched Surface of Limestone	123
7-4	Photomicrograph in Plane-Polarized Light of Thin Section Cut Through Grapestone Clump	124
7-5	View of Oncolites on Polished Surface	125
7-6	Photomicrograph in Plane-Polarized Light of Thin Section Through Oöid having Radial Fabric	125
7-7	View of Polished Surface of Dolomitic Carbonate Rock Showing Cross Sections of Vadose Pisolites	126
7-8	Slightly Modified Folk Limestone Classification	129
7-9	Photomicrograph in Plane-Polarized Light of Thin Slice of Limestone Showing Neomorphic Crystals of Calcite	133
7-10	Photomicrograph in Plane-Polarized Light of Thin Slice of Limestone Showing Neomorphic Crystals of Calcite	134
7-11	Photomicrograph in Cross-Polarized Light of Thin Slice of Dolostone Showing Crystals of Two Sizes	134
7-12	Photomicrograph in Cross-Polarized Light of Thin Slice Showing Poikilotopic Gypsum	135
7-13	Photomicrograph in Plane-Polarized Light of Thin Slice of Limestone Showing Ghost of Molluscan Fragment	136
7-14	Photomicrograph in Plane-Polarized Light of Thin Slice of Dolostone Showing Xenotopic Fabric	136
7-15	Sketch of View of Thin Slice of Dolostone	139
7-16	Sketch of Crystalline Fabric in Sedimentary Rock	140
7-17	Photomicrograph in Plane-Polarized Light of Thin Slice of Dolostone Showing a Hypidiotopic Fabric	141
7-18	Photomicrograph in Cross-Polarized Light of Thin Slice of Chert	144
7-19	Acid-Etched Sample of Core from Borehole Showing Laths of Anhydrite Pseudomorphs after Gypsum	146
7-20	Photomicrograph in Plane-Polarized Light of Thin Slice of Anhydrite Rock	147
7-21	Photomicrograph in Plane-Polarized Light of Thin Slice of Phosphate Rock Showing Oöids Composed of Collophane	148

Figure		Page
7-22	Photograph of Exposure of "Banded-Iron Formation"	150
8-1	Triangular Diagram Showing the Composition of Various Kinds of Sandstones	163
8-2	Core from Borehold Showing Well-Rounded Chert Pebbles in Oligomictic Conglomerate	166
8-3	Photograph of Exposure of Diamictite (tillite) Showing Pebbles and Boulders, Mostly Gneisses, Embedded in Sandy and Argillaceous Matrix	167
9-1	Standard Stratigraphic Symbols	189

LIST OF TABLES

Table		Page
2-1	Particle Grade-Size Scale	25
2-2	Zingg Shape Classes	35
3-1	Settling Times According to the Wadell Modification of Stokes' Law	46
5-1	Some Heavy Minerals and Their Parent Rocks	87
5-2	Minerals which are Magnetic at 0.4, 0.8, and 1.5 Amperes Settings on the Frantz Isodynamic Magnetic Separator	91
5-3	Characteristics and Mode of Occurrence of Important Heavy Minerals	97-103
5-4	Provenance of Some Common Heavy Minerals	105
6-1	Carbonate and Sulfate Minerals Subject to Stain Analysis	111
7-1	Sedimentary Rock Groups	119
7-2	Leighton-Pendexter Limestone Classification	121
7-3	Dunham Limestone Classification	127
8-1	Classifying and Naming Sandstones	160
9-1	Sedimentary Environments Considered in Exercise 9	175
9-2	Summary of Characteristics of Alluvial Facies	177
9-3	Summary of Characteristics of Tidal Facies	178
9-4	Summary of Characteristics of Nearshore Facies	179
9-5	Summary of Characteristics of Offshore Facies	180
9-6	Section 1	182
9-7	Section 2	183
9-8	Section 3	184
9-9	Section 4	185
9-10	Section 5	186
9-11	Section 6	187
9-12	Section 7	188

LIST OF REPORT FORMS

Form		Page
1-1	Report on Microscopic Examination of Sedimentary Particles	9
1-2	Pie Diagrams	10
2-1	Report on Sieve Analysis	30
2-2	Report on Microscopic Examination of Sieved-Size Fractions - Mineralogical Composition	33
2-3	Pie Diagrams - Mineralogical Composition of Sieved-Size Fractions	34
2-4	Report on Microscopic Examination of Sieved-Size Fractions - Sphericity and Roundness	37
3-1	Pipette Analysis	47
4-1	Histogram - Results of Exercise 2 Sieve Analysis	57
4-2	Frequency-Distribution Curve - Results of Exercise 2 Sieve Analysis	61
4-3	Cumulative-Frequency Curve (Arithmetic Ordinate) - Results of Exercise 2 Sieve Analysis	63
4-4	Cumulative-Frequency Curve (Probability Ordinate) - Results of Exercise 2 Sieve Analysis	67
4-5	Statistical Parameters Computed from Cumulative-Frequency Curve - Results of Exercise 2 Sieve Analysis	70
5-1	Heavy-Mineral Separation by Gravity Method	92
5-2	Determination of Heavy-Mineral Composition	104
6-1	Stained Carbonate-Rock Peels - Observational Data on Texture, Fabric and Mineralogical Composition	116
7-1	Report on Classification of Intrabasinal Sedimentary Rocks	153-154
8-1	Report on Classification of Extrabasinal and Pyroclastic Rocks	171-172

EXERCISE 1

> The surface of the land is made by nature to decay ..... Our fertile plains are formed from the ruins of the mountains.

James Hutton, 1785

EXERCISE 1
SEDIMENTARY PARTICLES

Introduction

Sedimentary deposits are assemblages of individual grains, or particles. These particles compose the framework of a modern sand or mud deposit, such as a beach or a point bar, or of its lithified equivalent, a sedimentary rock. The purpose of this exercise is to introduce you to the various kinds of particles that are commonly found in sedimentary rocks. Depending on their kind, they may be excellent indicators of the nature of either (1) the source from which they were derived or (2) the environment in which they evolved and accumulated.

Kinds of Sedimentary Particles

Sedimentary particles may be divided into two broad groups (Friedman and Sanders, 1978): (1) solid breakdown products of older deposits and (2) materials that are not solid breakdown products of older deposits. The second group consists of (2a) materials violently ejected from volcanoes and (2b) materials transported in water solution and deposited at the site of accumulation by biologic secretion or chemical precipitation.

Group-One Particles: Solid breakdown products of older deposits. All particles which are eroded from the land constitute terrigenous particles. These are subdivided into two groups: (a) inorganic materials and (b) carbonaceous materials. Most terrigenous particles, if traced back to their original place of origin, come from bedrock. Examples of inorganic terrigenous particles are rock fragments, quartz, feldspar and heavy minerals. Heavy minerals, as the name implies, are more dense than most minerals. They were originally derived from igneous or metamorphic rocks. Carbonaceous terrigenous particles include: (1) solid carbonaceous materials reworked from older strata and (2) modern plant detritus.

Group-Two Particles: Materials that are not solid breakdown products of older deposits. These are subdivided into: (a) materials ejected

Bullshis

from volcanoes and (b) solids that grew biochemically or chemically in the depositional basin. Materials ejected from volcanoes are considered pyroclastic particles. They may be rock fragments (lithic), single crystals (crystal), or bits of volcanic glass (vitric). Accumulations of pyroclastic particles are sometimes referred to as tephra. Particles that grew as solids in the depositional basin include (1) skeletal debris, (2) nonskeletal calcium-carbonate materials, (3) evaporite minerals that were physically transported and (4) glauconites.

References Dealing with Sedimentary Particles

Bathurst, R. G. C., 1971, Carbonate sediments and their diagenesis: Amsterdam, Elsevier Publishing Co., 620 p., Chapter 2. (Petrography of carbonate grains, 2: Oöids, pisolites, peloids, and other micritic fabrics, p. 77-91.)

Blatt, Harvey, 1967a, Original characteristics of clastic quartz grains: Jour. Sedimentary Petrology, v. 37, p. 401-424.

__________, 1967b, Provenance determinations and recycling of sediments: Jour. Sedimentary Petrology, v. 37, p. 1031-1044.

__________; Middletown, G. V.; and Murray, R. C., 1980, Origin of sedimentary rocks, 2nd ed.: Englewood Cliffs, NJ, Prentice-Hall, 634 p., Chapter 8. (Mineral composition of sandstones, p. 280-331.)

Boggs, Sam, Jr., 1968, Experimental study of rock fragments: Jour. Sedimentary Petrology, v. 38, p. 1326-1339.

Friedman, G. M.; and Sanders, J. E., 1978, Principles of sedimentology: New York, NY, John Wiley & Sons, 792 p., Chapters 2 and 3. (Kinds of sedimentary particles, p. 25-57, and Properties of sedimentary particles, p. 58-81.)

Keller, W. D., 1970, Environmental aspects of clay minerals: Jour. Sedimentary Petrology, v. 40, p. 788-813.

Logvinenko, N. V., 1982, Origin of glauconite in the Recent bottom sediments of the oceans: Sedimentary Geology, v. 31, p. 43-48.

McRae, S. G., 1972, Glauconite: Earth Science Reviews, v. 8, p. 397-440.

Pettijohn, F. J.; Potter, P. E.; and Siever, Raymond, 1972, Sand and sandstone: New York-Heidelberg-Berlin, Springer-Verlag, 618 p., Chapter 7. (The volcaniclastics, p. 261-292.)

Pittman, E. D., 1970, Plagioclase feldspar as an indicator of provenance in sedimentary rocks: Jour. Sedimentary Petrology, v. 40, p. 591-598.

Simone, Lucia, 1981. Oöids: A Review: Earth-Science Reviews, Vol. 16, p. 319-355.

Tyler, S. A., 1940, Zircon studies in the New Jersey Highlands: Am. Jour. Sci., v. 238, p. 260-271.

Equipment and Materials

Binocular microscope
Metal tray (approximately 8×10 cm)
Watch glass
Tweezers
Mounted needle
Camel-hair brush
Hydrofluoric acid
Ventilated hood
Sodium-cobaltinitrite solution
Electric furnace
Distilled water
Plastic squeeze bottles
Alizarine red S
Feigl's solution

Procedures

This exercise consists of two parts: first, examination of particles derived from the breakdown of igneous, metamorphic and sedimentary parent rocks and, second, examination of mud, sand, gravel and calcium-carbonate particles collected from modern sedimentary environments.

For the first part, four samples are provided: crushed granite, crushed mica schist, crushed rhyolite and crushed sandstone. The samples consist of particles derived from crushing rock with a mortar and pestle and then sieving the crushed material through an 80-mesh sieve. Particles finer than 177 microns were rejected and the larger material retained for study. The samples are very similar in texture and composition to the material resulting from mechanical weathering (disintegration) of rocks in high latitudes or at high elevations. In lower latitudes and at lower elevations much of the weathering of rock is chemical (decomposition) and the weathered material tends to differ from your samples in that the particles would be less angular and, depending on the intensity of chemical modification, the composition of the material might be quite different.

Your crushed granite, schist and rhyolite samples are a good representation of first-cycle sediments; that is, material derived from weathering and erosion of an igneous and/or metamorphic source terrane.

Your crushed sandstone sample is representative of a second-cycle sedimentary deposit; that is, material derived from an area where sedimentary rocks are exposed. It is possible for sedimentary particles to go through several cycles of weathering, erosion, transportation and deposition, and particles, such as these, commonly have characteristics which permit the geologist to reconstruct their multicyclic history.

Single-mineral particles derived from the weathering of a granite or gneiss are quartz, feldspar and ferromagnesian minerals, such as pyroxene, amphibole or mica. Some particles, which contain more than one of these minerals and which can be recognized as fragments of the parent rock, are designated rock fragments. Although rhyolite has the same mineralogical composition as granite, it has a microcrystalline texture, and for this reason all particles derived from weathering of a rhyolite are, in fact, rock fragments. Mechanical breakdown of mica schist, inasmuch as one of the component minerals (mica) is very soft, will yield a mixture of particles quite different from that of the granite or rhyolite.

In your study of the mineralogical composition of the crushed granite sample, and perhaps the crushed sandstone sample, you may find that it is difficult to distinguish quartz particles from feldspar particles. If the quartz luster is glassy and if the feldspar luster is porcelaneous you will have little difficulty. Even if the feldspar happens to be glassy, if it exhibits cleavage and perhaps twinning you will still have no problem with quartz/feldspar differentiation. Some sediments exist, however, in which quartz and feldspar are so much alike that it is virtually impossible to distinguish between the two kinds of particles. For such sediments it is necessary to stain the feldspar.

Procedure I: Staining Feldspar in Sediments

1. For your crushed granite sample and crushed sandstone sample, etch the sediment particles in hydrofluoric acid (35%) vapor for one minute at 90°C. The particles must be in the vapor phase above the liquid hydrofluoric acid. THIS PROCEDURE SHOULD BE CARRIED OUT IN A WELL-VENTILATED HOOD. Since hydrofluoric acid may cause painful burns, extreme care should be taken in using this reagent. Gloves are advisable when handling it. FOR STUDENTS WHO ARE INEXPERIENCED IN CHEMICAL LABORATORY PROCEDURES, IT IS PREFERABLE FOR THE INSTRUCTOR TO ETCH SAMPLES FOR THE STUDENTS BEFORE THE LABORATORY PERIOD. STUDENTS, HOWEVER, SHOULD DO THE STAINING.

2. After etching, heat the samples in an electric furnace for about five minutes at 400°C.

3. Pour sodium-cobaltinitrite solution on the particles (1 g of sodium-cobaltinitrite in 4 ml of distilled water) and leave for about one minute.

4. Wash the samples in distilled water. K-feldspar grains show a yellow stain.

References Dealing with Staining of Feldspars

Baily, E. H.; and Stevens, R. E., 1960, Selective staining of K-feldspar and plagioclase on rock slabs and thin sections: Am. Mineralogist, v. 45, p. 1020-1026.

Friedman, G. M., 1971, Staining, Chapter 22, p. 511-530, in Carver, R. E., ed., Procedures in sedimentary petrology: New York, Wiley-Interscience, 653 p. (Staining of feldspars, p. 522-527.)

van der Plas, Leendert, 1966, The identification of detrital feldspars: Amsterdam, Elsevier Pub. Co., 305 p.

Procedure II: Examination of Fragments of Igneous, Metamorphic and Sedimentary Rock

1. Place the crushed granite sample in the metal tray and, using reflected light, examine it (dry) under the binocular microscope. Then carry out a second examination by placing the sample on the watch glass and viewing it (wet) using reflected and transmitted light. Which procedure gives you the better view of the sedimentary particles?

2. Study the mineralogical composition of each of your four samples and estimate percentages of quartz, feldspar, ferromagnesian minerals and rock fragments. Figures 1-1 and 1-2 contain photomicrographs in which there are examples of some of the kinds of particles that you will find in your crushed-rock samples. Designate those particles that you cannot recognize as Unknown #1, #2, etc. Try to keep the unknowns to a minimum. For each sample, use Form 1-1 to record your findings and Form 1-2 to prepare a pie diagram.

The kinds and abundances of mineral particles that you will find in your mud, sand and gravel samples depend on: (1) availability in the source area, (2) mechanical durability during sediment transport and (3) chemical stability in the settings in which they were weathered, transported and deposited. One of the intriguing things about terrigenous sediments and rocks is that, although only a relatively few minerals are present in them as important constituents, these minerals occur in many different mixtures and arrangements. For example, in some sedimentary

A

Crushed granite - includes particles of quartz, feldspar and mica. The composition approximates the initial stage of sediment formation through disintegration of a granite outcrop.

B

Crushed mica schist - contains biotite, muscovite, quartz and feldspar particles.

Photomicrographs of Particles which are Solid Breakdown Products of Older Deposits

Figure 1-1

A

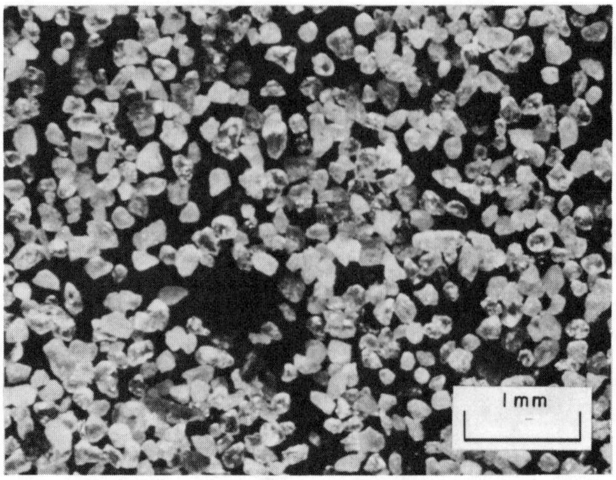

Modern quartz sediment from sea floor, 0.25 to 0.125 mm fraction, under binocular microscope (B. M. Shaub in Friedman and Sanders, 1978, Fig. 2-4, A, p. 30).

Crushed rhyolite - Crushed rhyolite rock fragments. Composition is the same as granite only with a finer grain size.

Photomicrographs of Particles which are Solid Breakdown Products of Older Deposits

Figure 1-2

REPORT ON MICROSCOPIC EXAMINATION
OF SEDIMENTARY PARTICLES

Exercise 1
Name ______________________
Laboratory Section ______________
Date ______________________

Sample Number	Nature of Sample (crushed rock, sand, soil, etc.)	Estimated Percentages of Various Kinds of Particles												Notes: Observable properties, identification of rock fragments, etc.)
		Terrigenous						Carbonate						
		Q	F	Fm	M	R	U	O_o	P	S	Lm	O	U	

Q = quartz F = feldspars Fm = ferromagnesians M = micas R = rock fragments
U = unknown O_o = oöids P = peloids S = skeletal debris Lm = lime mud (micrite) O = other

Form 1-1

PIE DIAGRAMS - proportions of particles in samples examined (3.6° of arc = 1% composition)

Exercise 1
Name ______________________________
Laboratory Section ______________
Date ______________________________

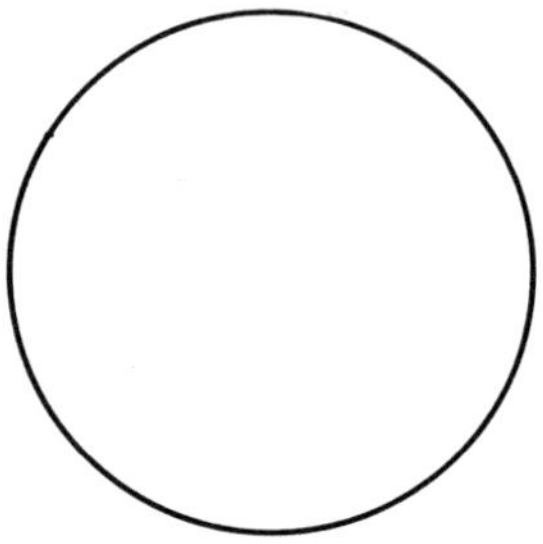

Crushed Granite

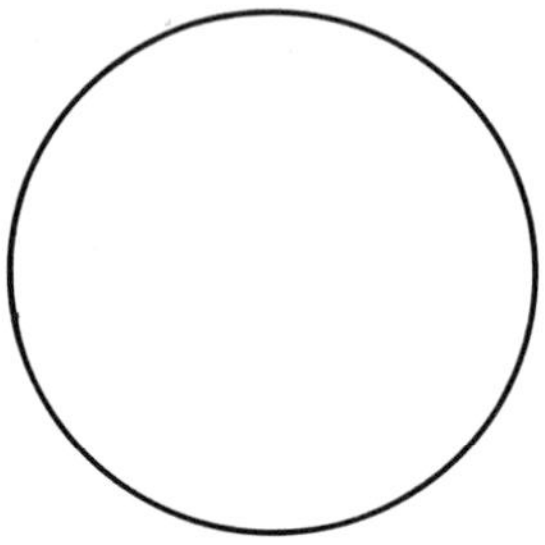

Crushed Rhyolite

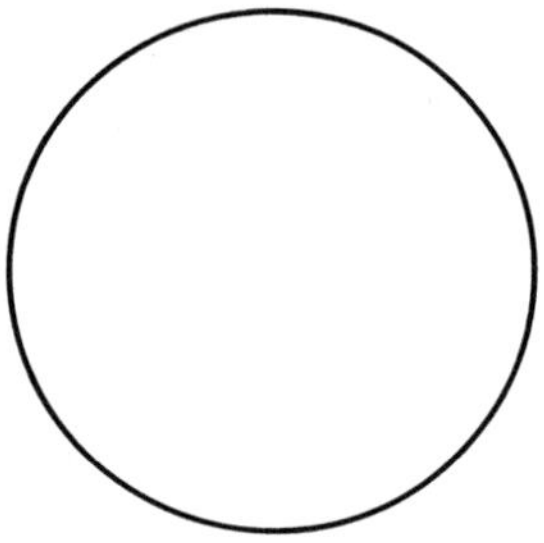

Crushed Mica Schist

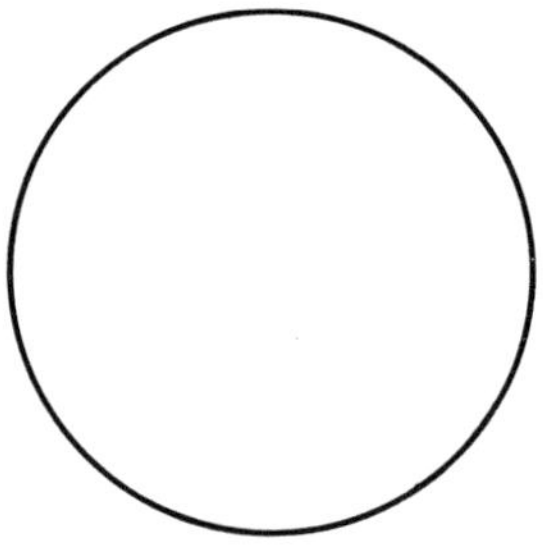

Crushed Sandstone

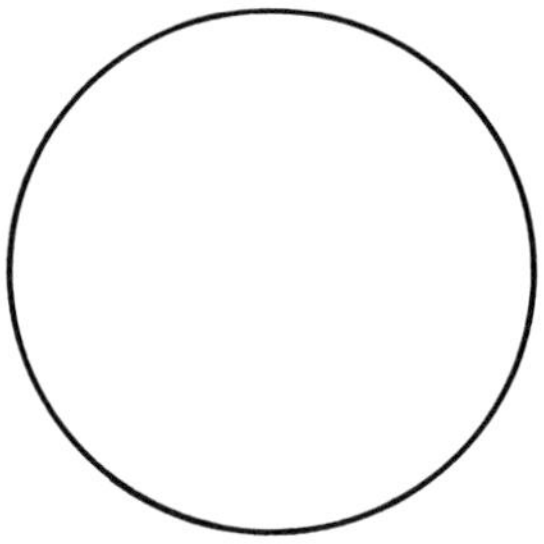

Mud

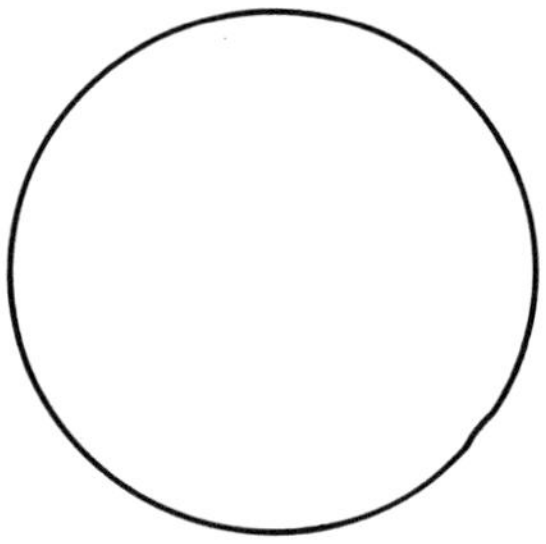

Sand

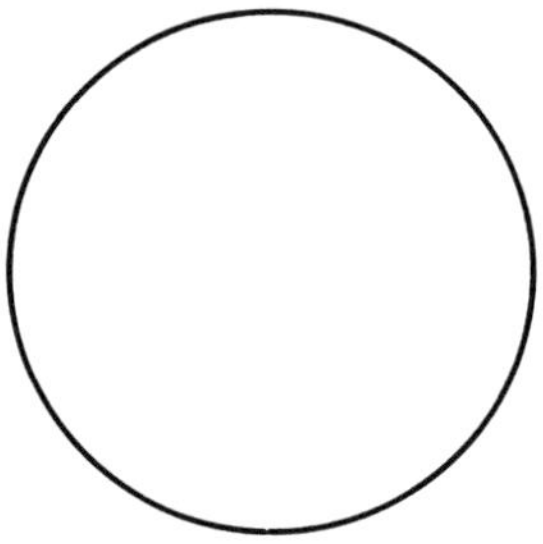

Gravel

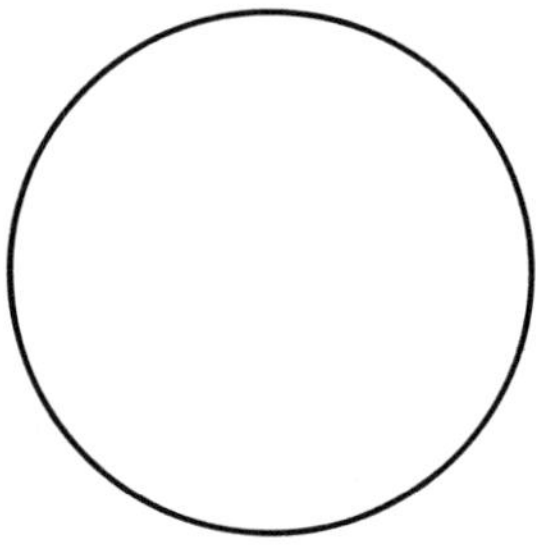

Carbonate #1

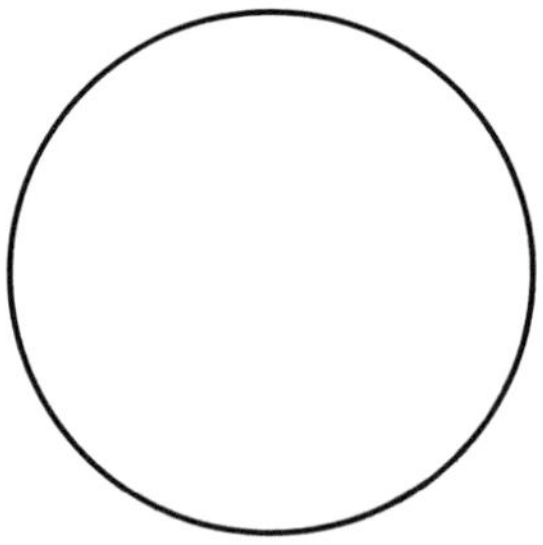

Carbonate #2

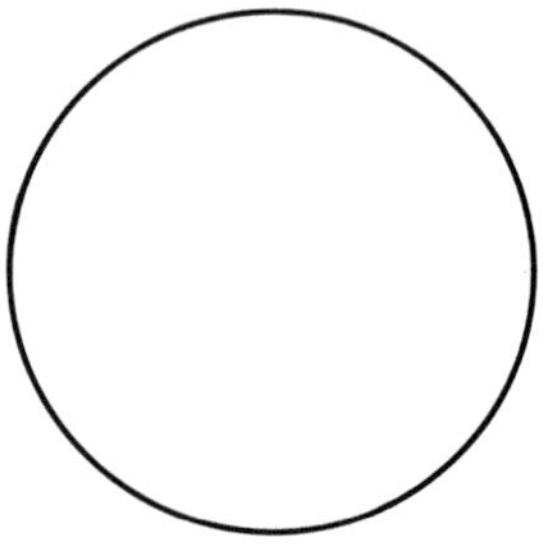

Carbonate #3

Form 1-2

rocks the mineral population may be largely unstable, first-cycle particles; in others it may consist largely of stable, second-cycle particles.

Figures 1-3 and 1-4 contain photomicrographs of some of the kinds of particles that you will find in your mud, sand and gravel samples.

Procedure III: Examination of Mud, Sand and Gravel

1. Inasmuch as some sands and gravels are rich in quartz and feldspar, it may be necessary for you to first stain these samples for feldspar, as you did in Procedure I: Staining Feldspar in Sediments.

2. Examine the mud, sand and gravel samples, both dry and wetted, under the binocular microscope using both transmitted and reflected light. To what extent does particle size effect your ability to determine mineralogical composition?

3. Record particle-type data on Form 1-1 and diagram your estimates of percentages of the various particle-types on Form 1-2.

So far, you have studied particles derived from crushing rock samples and particles in mud, sand and gravel collected from modern non-carbonate sedimentary environments. You will now study sedimentary particles collected from modern carbonate sedimentary environments.

Those samples containing carbonate particles are composed of material that grew as solids in the depositional basin, including oöids, pellets, peloids, skeletal debris and lime mud. Unlike evaporite sediments, which are precipitated out of solution under highly saline conditions in the depositional basin, these kinds of particles are generated by biological, chemical, or biochemical processes, moved about by physical forces and biological activity, and finally deposited.

Oöids derive their name from a Greek word that means egg or egglike, because these particles resemble fish roe. They are usually spherical or elliptical and generally consist of aragonite. Oöids have a concentric or radial internal structure. Pellets are formed by deposit-feeding organisms that eat the lime mud, digest included organic matter and excrete the non-digestible matter (usually aragonite) as sand-size spherical or ellipsoidal fecal pellets. These are usually internally homogeneous and devoid of structure. Peloids are sand-size, spherical or ellipsoidal particles of calcium carbonate for which no particular origin is implied.

A

Gravel - Rock fragments, the first stage of particle generation; minerals have not yet been separated from rock matrix.

B

Garnetiferous Sand - Concentration of resistant minerals including garnet.

Photomicrographs of Particles
Which are Solid Breakdown Products of Older Deposits

Figure 1-3

A

Hawaiian Green Sand - Heavy minerals including magnetite, rutile and ilmenite and other ferromagnesium minerals concentrated in sand from pyroclastic debris. The green color of the sand is caused by olivine.

B

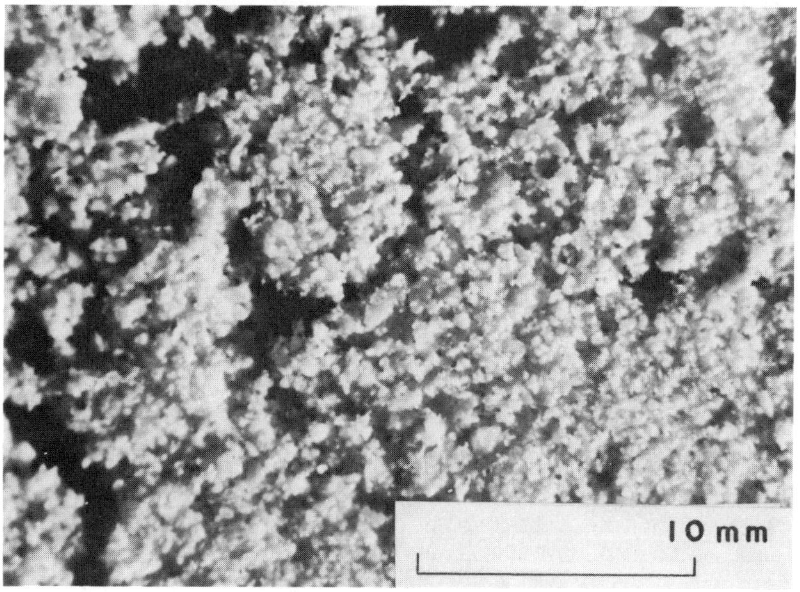

Clay - View of clay through binocular microscope.

Photomicrographs of Particles which are
Solid Breakdown Products of Older Deposits

Figure 1-4

Most skeletal materials secreted by the metabolic activities of living organisms are composed of calcium carbonate or silica. Upon death of the organism the skeletal structure is fragmented and, depending on size, particles of skeletal debris in aggregate form gravel, sand or mud. The particles may be derived from such sources as reef-rock, bivalve material, gastropod shells, coccolithophore plates, sea-urchin spines and bryozoans. Most lime mud consists of tiny needles and platelets of carbonate crystals too small to resolve by binocular or petrographic microscopes. Almost all carbonate particles composing lime mud are derived from organism secretion of tiny particles, by accumulation of tiny skeletal components secreted by algae or by physical breakdown of sand-size and larger skeletal material.

Figures 1-5 and 1-6 contain photographs of carbonate particles derived from material that grew as solids in the depositional basin.

Procedure IV: Staining Calcite and Dolomite in Sediments

1. Wash sample under running water and etch with dilute hydrochloric acid for two to three minutes (acid solution consists of eight to ten parts by volume of commercial grade, concentrated HCl diluted with water to 100 parts).

2. Rinse sample with water; dry sample.

3. Drop alizarine red S on sample and allow staining action to take effect for one to three minutes (alizarine red S is prepared by dissolving 0.1 g of alizarine red in 100 ml of 0.2% HCl).

4. Gently rinse sample with water and examine sample under binocular microscope.

5. Those carbonate particles which stain pale red are calcite or aragonite; those which do not are dolomite.

6. To distinguish calcite from aragonite it is necessary to stain the sample using Feigl's solution, a test which depends on the slightly different solubilities of aragonite and calcite in water. (The method of preparing Feigl's solution is described in Katz and Friedman, 1965.)

More detailed information on staining techniques, especially as they relate to carbonate rocks, is included in Exercise 6 of this manual, CARBONATE STAINING AND PEELS.

A

Oöids from Abu Dhabi, Trucial Coast, Persian Gulf

B

Skeletal debris, bedding surface of limestone (Friedman and Sanders, 1978, Fig. 2-19, p. 47).

Photographs of Carbonate Particles
that Grew as Solids in the Depositional Basin

Figure 1-5

Carbonate debris - Carbonate particles including a mixture of oöids, grapestones, pellets, and skeletal debris.

Photomicrograph of Carbonate Particles
that Grew as Solids in the Depositional Basin

Figure 1-6

References Dealing with Staining of Carbonates

Friedman, G. M., 1971, Staining, Chapter 22, p. 511-530, in Carver, R. E., ed., Procedures in sedimentary petrology: New York, Wiley-Interscience, 653 p. (Staining of carbonates, p. 513-519.)

Friedman, G. M., 1977, Identification by staining methods of minerals in carbonate rocks, p. 96-97, in LeRoy, L. W.; LeRoy, D. D.; and Raese, J. W., eds., Subsurface geology, 4th ed.: Golden, Colorado, Colorado School of Mines, 941 p.

Katz, Amitai; and Friedman, G. M., 1965, The preparation of stained acetate peels for the study of carbonate rocks: Jour. Sedimentary Petrology, v. 35, p. 248-249.

Warne, Slade, 1962, A quick field or laboratory staining scheme for the differentiation of the major carbonate minerals: Jour. Sedimentary Petrology, v. 32, p. 29-38.

Wolf, K. H.; and Warne, Slade, 1960, Remarks on the application of Friedman's staining methods: Jour. Sedimentary Petrology, v. 30, p. 496-497.

Procedure V: Examination of Samples Containing Carbonate Particles

1. Examine carbonate samples, both wetted and dry, under the binocular microscope using both transmitted and reflected light.

2. Determine the composition of each sample from the standpoint of both mineralogy and particle mode of origin (i.e., distribution within the sample of oöids, skeletal fragments, peloids, etc.).

3. Record particle-type data on Form 1-1 and diagram on Form 1-2 your estimates of percentages of the various particle types.

References Dealing with the Identification of Carbonate Sedimentary Particles

Bathurst, R. G. C., 1976, Carbonate sediments and their diagenesis: Amsterdam, Elsevier Scientific Publishing Co., 658 p.

Ham, W. E., ed., 1962, Classification of carbonate rocks: Tulsa, American Assoc. Petroleum Geologists, 279 p.

Horowitz, A. S.; and Potter, P. E., 1971, Introductory petrography of fossils: Berlin, Springer-Verlag, 302 p.

Scholle, P. A., 1978, A color illustrated guide to carbonate rock constituents, textures, cements, and porosities: Tulsa, American Assoc. Petroluem Geologists, 241 p.

You will gain additional experience in the identification of carbonate particles when you do Exercise 6 - CARBONATE STAINING AND PEELS and Exercise 7 - DESCRIPTION AND CLASSIFICATION OF SEDIMENTARY ROCKS IN HAND SPECIMEN AND THIN SECTION: INTRABASINAL ROCKS.

Other Examples of Group-Two Particles

In addition to carbonate particles, evaporite minerals and glauconite are also examples of particles that grew as solids in the depositional basin. Glauconite particles are common in some marine rock strata and sediments. These particles appear under the binocular microscope as sand-size, green, earthy pellets. The shapes of such pellets vary; they may be ovoidal, spherical (Figure 1-7) or botryoidal; they may fill chambers of foraminifers and preserve the internal shapes of the foraminifers as molds when the calcite tests have been dissolved. Commonly, however, the shapes of these green pellets defy precise description.

Mineralogists recognize four groups of minerals that occur as small, round, green pellets: (1) pellets that have the structural properties of glauconite, with three x-ray diffraction peaks, (2) pellets that likewise have the structural properties of the mineral glauconite, but with subdued, broad asymmetric x-ray peaks, (3) pellets that consist of interlayered clay minerals and (4) pellets that consist of a mixture of clay minerals.

Glauconite occurs in minor amounts in marine sediments. In places, such as along the coasts of New Jersey and southeastern England, glauconite is abundant because it is being eroded offshore from older glauconite-rich strata. These sediments are referred to as "greensands." An excellent glauconite reference is McRae, S. G., 1972, Glauconite: Earth-Science Reviews, v. 8, p. 397-440.

Pyroclastic particles, which are derived explosively from volcanoes, can include rock fragments, single crystals or bits of volcanic glass. Rock fragments can consist of previously solidified volcanic rocks

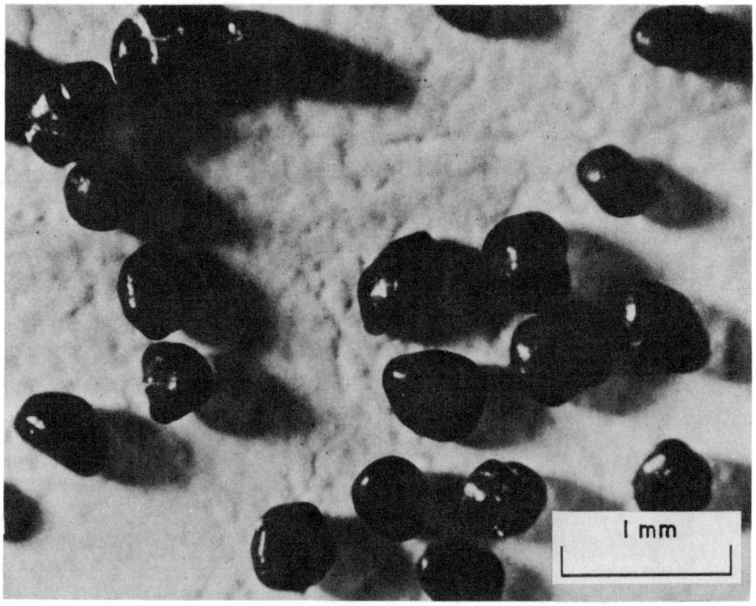

Spherical and ovoidal pellets of glauconite, dark green in color, under binocular microscope. Dark green color shows here as black. (D. M. Triplehorn, 1966, Fig. 1, p. 249.)

Figure 1-7

or of any kind of rock through which the volcanic gases and the lava passed on their way to the surface. The crystals must have already grown within the magma. The glass particles represent blebs of molten lava that solidified so rapidly that their ions did not form crystal lattices.

As noted previously, certain features characterize mineral particles from volcanic sources: doubly terminated prisms of β-quartz, rounded and embayed quartz particles, oscillatory zoning in plagioclase, and bleblike inclusions within quartz or feldspar. Typical tephra consists of shards of glass having distinctive shapes indicating that they broke from hollow spheres, which formed as the gas frothed through the hot lava (Fig. 1-8). Tephra generally are so distinctive that they are readily recognized. The use of tephra in provenance studies requires that petrographic analysis be supplemented by determinations of the index of refraction of glass particles and analysis of trace elements.

Tephra have been spread widely from some spectacular volcanic explosions. If the tephra can be traced to their parent volcano, and if they were discharged at a known date, then they can serve as valuable time markers. An example of the use of tephra layers as time markers is the

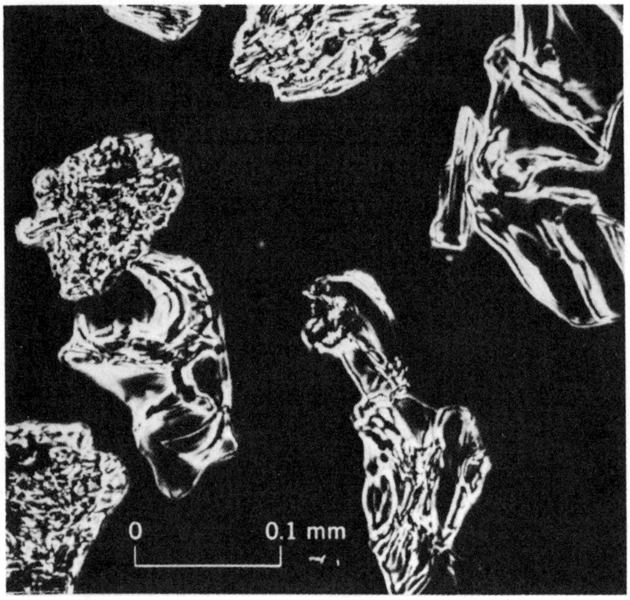

Photomicrograph of glass shards of tephra from explosive eruption of Mount Mazama (ancestral peak at Crater Lake Oregon). Quaternary sediments, Creston Bog, Washington. (R. B. Taylor, in Friedman and Sanders, 1978, Fig. 2-13, p. 44.)

Figure 1-8

anaylsis of cores of sediment from the floor of the eastern Mediterranean Sea. Many cores contain two layers of tephra from the Volcano Santorini. The older layer is estimated to be about 25,000 years old, whereas the younger layer dates at about 1450 B.C., a date that coincides with the abrupt demise of the Minoan civilization on the island of Crete. Figure 1-9 is a view of tephra that evolved during a recent eruption of a volcano in the northwestern United States.

Photomicrograph of volcanic ash from the Spring, 1981 violent eruption of Mount St. Helens, Washington. Elongate particles are fragments of conifer needles.

Figure 1-9

EXERCISE 2

An everywhere of silver
With ropes of sand
To keep it from effacing
The tract called land.

Emily Dickinson

EXERCISE 2
TEXTURAL ANALYSIS OF SAND

Introduction

In exercise 1 we examined samples containing a variety of terrigenous sedimentary particles. Such study of mineralogical composition yields useful information about the source terrane from which the particles were derived. Careful study of sediment texture yields important information about the processes which were operative in transporting and depositing particles. The textural attributes of a sedimentary deposit or rock include particle (1) size, (2) shape, (3) surface texture, and (4) arrangement (packing and fabric). Exercise 2 is designed to provide an introduction to analysis of sand texture. It consists of two parts: (1) determination of the size distribution of particles in sand samples and (2) study of component particle shapes and surface textures. The textural analysis of finer particles, which require different techniques, will be treated in Exercise 3.

Particle-Size Analysis

Sediment particle-size is important in a number of ways. It reflects (1) the availability of different sizes of particles in the bedrock or pre-existing sediments of the source area, (2) the resistance of particles of various compositions to breakdown during weathering and transportation and (3) the kinds of processes responsible for transportation and deposition of the sediment.

Although sedimentary particles occur in nature as a continuum ranging from fine clay to very large boulders, geologists have chosen to devise grade scales with size classes having specific lower and upper limits. A commonly used grade scale, modified from C. K. Wentworth (1922), is shown in Table 2-1. Note that on this scale particle diameters are expressed in millimeters and that size classes are based on an octave scale; that is, adjacent classes differ by a factor of two. In the mid-1930's W. C. Krumbein developed the phi scale of particle-grade size as a device to permit the direct application of conventional statistical practices to sedimentary data. He accomplished this by using the

Limiting Particle Diameter (mm)	(φ units)	(Microns μ)	Size Class			
			V. Large	Boulders	GRAVEL	
2048	-11		Large	Boulders	GRAVEL	
1024	-10		Medium	Boulders	GRAVEL	1m
512	- 9		Small	Boulders	GRAVEL	
256	- 8		Large	Cobbles	GRAVEL	
128	- 7		Small	Cobbles	GRAVEL	10^{-1}
64	- 6		V. Coarse	Pebbles	GRAVEL	
32	- 5		Coarse	Pebbles	GRAVEL	
16	- 4		Medium	Pebbles	GRAVEL	
8	- 3		Fine	Pebbles	GRAVEL	10^{-2}
4	- 2		V. Fine	Pebbles	GRAVEL	
2	- 1		V. Coarse	Sand		
1	0		Coarse	Sand		10^{-3}
1/2	+ 1	500	Medium	Sand		
1/4	+ 2	250	Fine	Sand		
1/8	+ 3	125	V. Fine	Sand		10^{-4}
1/16	+ 4	62	V. Coarse	Silt	MUD	
1/32	+ 5	31	Coarse	Silt	MUD	
1/64	+ 6	16	Medium	Silt	MUD	
1/128	+ 7	8	Fine	Silt	MUD	10^{-5}
1/256	+ 8	4	V. Fine	Silt	MUD	
1/512	+ 9	2		Clay	MUD	

what kind of crap is this no decimal numbers?

Particle Grade-Size Scale
(Friedman and Sanders, 1978)

Table 2-1

transformation equation $\phi = -\log_2 d$, where d = particle diameter in mm. Phi values, which are dimensionless numbers that increase with decreasing particle size, are shown with their millimeter equivalents in Table 2-1. The principal advantages of using the phi scale are that (1) the distribution of particle sizes can be plotted easily on arithmetic graph paper, (2) calculation of various statistical parameters is simplified, (3) limiting particle diameters for size classes are whole numbers rather than unwieldy fractions or two or three digit numbers, and (4) negative log usage permits a more convenient and consistent method of plotting data. Exercise 4 will deal with graphic presentation of particle-size data and the statistical parameters of particle-size frequency distribution.

References Dealing with Sediment Particle Size Analysis

Folk, R. L., 1966, A review of grain-size parameters: Sedimentology, v. 6, p. 73-93.

Friedman, G. M., 1961, Distinction between (sic) dune, beach and river sands from their textural characteristics: Jour. Sedimentary Petrology, v. 31, p. 514-529.

__________, 1967, Dynamic processes and statistical parameters compared for size frequency distribution of beach and river sands: Jour. Sedimentary Petrology, v. 37, p. 327-354.

__________, 1979, Differences in size distributions of populations of particles among sands of various origins: Sedimentology, v. 26, p. 3-32.

__________, 1979, Differences in size distributions of populations of particles among sands of various origins: Addendum to IAS presidential address: Sedimentology, v. 26, p. 859-862.

__________; and Sanders, J. E., 1978, Principles of sedimentology: New York, NY, John Wiley & Sons, 792 p., Chapter 3. (Properties of sedimentary particles, p. 58-81.)

Krinsley, D. H.; and Smalley, I. J., 1972, Sand: Am. Scientist, v. 60, p. 286-291.

Krumbein, W. C., 1934, Size frequency distribution of sediments: Jour. Sedimentary Petrology, v. 4, p. 65-77.

__________, 1936, The application of logarithmic moments to size frequency distributions of sediments: Jour. Sedimentary Petrology, v. 6, p. 35-47.

__________; and Pettijohn, F. J., 1938, Manual of sedimentary petrography: New York, NY, Appleton-Century-Crofts, Inc., 549 p., Chapters 4-7. (Concept of a grade scale, p. 76-90; Principles of mechanical analysis, p. 91-134; Methods of mechanical analysis, p. 135-181; Graphic presentation of analytical data, p. 182-211.)

Wentworth, C. K., 1922, A scale of grade and class terms for clastic sediments: Jour. Geology, v. 30, p. 377-392.

Equipment and Materials

Knife or spatula with large blade
Sheets of high-gloss paper (approx. 40 x 40 cm)

Mechanical sample splitter
Balance
Wire-mesh sieves
Sieve shaker
Brush for cleaning sieves
Binocular microscope

Determination of Particle-Size Distribution

The first step in carrying out textural analysis of a sediment is to split the sample. The purpose of splitting a sample is (1) to obtain material representative of the sample originally taken in the field and (2) to reduce the sample size to that required in sieving analysis (25 g).

Two convenient splitting methods are commonly used: (1) splitting by hand, and (2) splitting by a mechanical sample splitter (Fig. 2-1). In this exercise you will use both methods.

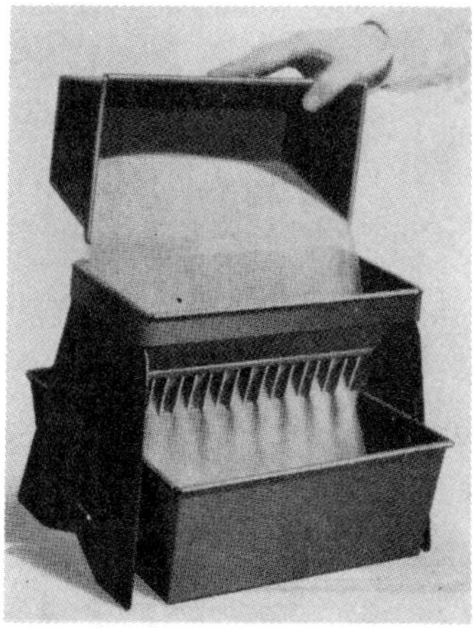

Mechanical Sample Splitter

Figure 2-1

Procedure I: Splitting the Sample

1. Heap 200 g of sand into a cone on a glossy sheet of paper.
2. Cut the cone into quarters with a knife or spatula; reject the first and third or second and fourth quarters.

3. Heap the remaining two quarters into a cone again, and divide the sample once more into quarters; reject quarters as before. You should have a sample weighing about 50 g. (If for some reason you started with more than 200 g, repeat this division until you reach the desired sample size.)

4. The 50 g sample should now be split with a mechanical splitter. The splitter consists of chutes inclined alternately in opposite directions. Your sand sample will be divided into two halves, each consisting of about 25 g collected in a separate pan. Discard one of the two 25 g splits and save the other one for Procedure II.

Sieving is a method of particle-size separation by screening. It became important in detailed analyses of sediments after the 15th Century, when woven wire screens were introduced in Germany. It is now possible to separate particle sizes down to as fine as 25 microns, using woven wire sieves. Sieving is not practical, however, for sizes below approximately 62 microns (1/16 mm) due to electrostatic attraction of particles.

The equipment that you will need for sieving includes (1) a balance for weighing samples, (2) a set of nested wire-mesh sieves for separation of the various particle-grade sizes, and (3) a sieve-shaker apparatus to speed the passage of particles through the column of sieves. Some laboratories use a sonic shaker apparatus which sieves particles rapidly and at the same time permits observation of the movement of the particles from sieve to sieve. An example of this type of device is the Allen-Bradley Sonic Sifter (Fig. 2-2). Another type of sifting device, the Tyler Ro-Tap Testing Sieve Shaker (Fig. 2-2), in addition to being very noisy, requires more time for the sieving operation than the sonic devices. Of greater importance in terms of speeding the sieving operation is the choice of the balance to be used for weighing samples. Use of the triple-beam balance can be time consuming and for this reason an electronic balance is recommended (Fig. 2-3).

Procedure II: Sieving the Sample

1. Weigh sand sample to nearest 0.01 g and record weight on Form 2-1. Your instructor will indicate an appropriate approximate weight for your sample. 50g

2. Pour sample into uppermost (coarsest) sieve and sieve for two minutes if using a sonic shaker or for 10 minutes if using a Ro-Tap shaker. Your instructor will recommend the appropriate sieve sizes for your sample.

Electronic Balance

Figure 2-3

3. Remove nested sieves from shaker and carefully pour contents of each sieve onto a sheet of clean, high-gloss paper (which reduces loss by adhesion due to static electricity).

4. Clean sieve screen by quickly inverting it and slamming it sharply and evenly on the paper. Be sure that table top is even and that the screen is not damaged by the procedure. Screen can be cleaned further by brushing it gently with fine to medium-bristle paint brush while it is inverted on paper. NEVER TOUCH THE SCREEN WITH ANYTHING OTHER THAN THE BRUSH!

5. Weigh fraction retained on each sieve and record weight on Form 2-1.

6. Save each fraction for later use by wrapping it in a fully labelled piece of paper. The "fines" (particles < 1/16 mm) may be used for Exercise 3.

7. Determine sieve loss (weight and weight percent of material lost in the sieving procedure) and record on Form 2-1.

Relationship of Particle Size to Mineralogical Composition

In some sands and sandstones there are striking mineralogic variations that exist as a function of particle size. For example, in a given sand sample one may find that the finer fractions are relatively feldspar-rich and quartz-poor and that the coarser fractions are quartz-rich and feldspar-poor. Another sample may have just the reverse relationship in terms of composition and grain-size. Your purpose in the next procedure will be to look for significant compositional differences as you compare the various sieved size fractions.

Procedure III: Study of Mineralogical Composition of Individual Size Fractions

1. Study each size fraction under the binocular microscope.
2. Using Form 2-2, record the mineralogical composition of each fraction as you did for sedimentary particles in Exercise 1.
3. Using Form 2-3, construct a pie diagram for each size fraction showing percentage abundances of particle types in the fraction.
4. In the Notes and Remarks column of Form 2-2, summarize your principal observations on significant compositional differences in the size fractions as you trace them from coarsest to finest.

Study of Particle Shapes and Surface Textures

The shape of a sedimentary particle reflects its origin, history and internal lattice structure. Particles precipitated by organisms vary widely in form and range from simple to exceedingly complex. For example, coccolithophores secrete disks, fusulinids are spindle-shaped, crinoid stems or ossicles are button-shaped and sea urchins release conical spines. Terrigenous particles released by weathering of bedrock also have many shapes that reflect their origins. Among these are euhedral quartz particles released from limestone beds, glass shards of tephra and very large, euhedral to subhedral quartz crystals weathered from some volcanic rock terranes.

As sedimentary particles are transported away from their place of origin their original shapes are modified as they collide with one another or with bedrock. Thus an initially angular particle may become round as edges and corners are worn away. Shapes of particles modified by abrasion during transport depend on (1) initial shape as liberated from bedrock, (2) composition - whether a particle consists of one or several minerals or rock fragments, (3) hardness, brittleness or toughness, (4) inherited partings, such as fractures, joints, bedding, schistosity or cleavage, (5) size, (6) agent of transport, (7) rigors of transport, including distance and energy of the agent of transport, and (8) other random effects. Particle shapes are expressed in terms of varying sphericity and roundness.

Sphericity defines the degree to which a particle approximates the shape of a sphere; that is, it expresses how nearly equal the three mutually perpendicular dimensions of a particle are. Sphericity has been defined as a function of (1) the surface area of the particle and the surface area of a sphere having a volume equal to that of the particle, (2) the volume of the particle and the volume of the smallest sphere that encloses the particle, and (3) the long, intermediate and short diameters of the particle. Since sphericity is an indication of how nearly equal the dimensions of a sand particle are, it provides an indication of the original shape of the particle. The parameter reflects the shape of the particles or crystals in the parent rock rather than the erosional or depositional history of the particle. Many techniques for measuring sphericity have been devised, the most straightforward being that of Zingg (1935).

The Zingg technique involves comparing the length (a), breadth (b) and thickness (c) of sand-size particles, as observed under the binocular microscope. You will do this by estimation rather than by exact measurement using a calibrated microscope ocular. Table 2-2 outlines the defining characteristics of the various Zingg shape classes.

Class Number	b/a	c/b	Shape
I	$>2/3$	$<2/3$	Oblate (discoidal)
II	$>2/3$	$>2/3$	Equiaxial (spherical)
III	$<2/3$	$<2/3$	Triaxial (bladed)
IV	$<2/3$	$>2/3$	Prolate (rod-shaped)

a = length; b = breadth; c = thickness

Zingg Shape Classes
Table 2-2

Procedure IV: Estimation of Sphericity

1. Using the coarsest size fraction from your sieved material and also one of the "middle" size fractions, randomly select 20 particles from each and estimate the Zingg b/a and c/b ratios for each particle.

2. Record observations on Form 2-4 and determine Zingg class numbers.

3. Compare the values observed for the two different size fractions.

References Dealing with Particle Sphericity

Brewer, Roy, 1964, Fabric and mineral analysis of soils: New York, NY, John Wiley & Sons, 470 p.

Curray, J. R.; and Griffiths, J. C., 1955, Sphericity and roundness of quartz grains in sediments: Bull. Geol. Soc. America, v. 66, p. 1075-1096.

Friedman, G. M.; and Sanders, J. E., 1978, Principles of sedimentology: New York, NY, John Wiley & Sons, 792 p., Chapter 3, Properties of sedimentary particles. (Sphericity, p. 60-61.)

Krumbein, W. C.; and Pettijohn, F. J., 1938, Manual of sedimentary petrography: New York, NY, Appleton-Century-Crofts, Inc., 549 p., Chapter 11. (Shape and roundness, p. 227-302.)

Zingg, Th., 1935, Beitrag zur Schotteranalyse: Schweiz. Mineralog. Petrog. Mitt., Bd. 15, p. 39-140.

Roundness is an attribute of particle shape that is related to the sharpness or curvature of edges and corners. It is geometrically independent of sphericity and reflects, in part, the abrasional history of the particle. Roundness is defined by the relationship among the radii of individual edges and corners, the number of corners measured and the radius of the maximum inscribed circle of the particle. The roundness of sand-size particles can most easily be determined by matching the outlines of individual particles with diagrams of two sets of standard sand-size particles, each set having different sphericity (Fig. 2-4). This is the technique that we will use in Procedure V.

... do you not see that stones
even are conquered by time, that
tall turrets do fall and rocks
do crumble?

Lucretius

EXERCISE 3
SIZE ANALYSIS OF SILT AND CLAY

Introduction

A non-consolidated sedimentary deposit consisting of particles having diameters less than 4ϕ is considered a mud. Stated more elegantly, the deposit's component particles fall in the silt- and clay- grade size range (Table 2-1). Also, some sand deposits contain a significant admixture of silt and clay particles.

If the fine material in a sand sample is predominantly silt, the entire sample is dried, disaggregated, weighed and dry-sieved (the procedure that you followed in Exercise 2). The material collected in the pan, called fines, is then analyzed separately from the sand fraction using the procedures that you will follow in this exercise. If the fine material in a sand sample is predominantly clay, dry-sieving cannot be used because the extremely fine clay particles will clog the screens; wet-sieving or decantation procedures must be used.

The two most widely used methods of size analysis are the pipette and the hydrometer techniques. Both of these involve "sedimentation" in which particle sizes and quantities are estimated from the rates at which the particles sink in a fluid medium. In this exercise you will conduct a pipette analysis. The hydrometer technique, although rapid, is criticized by some researchers as not being sufficiently accurate for most sedimentologic investigations.

References Dealing with Size Analysis of Silt and Clay

Folk, R. L., 1974, Petrology of sedimentary rocks: Austin, Hemphill Publishing Co., 182 p.

Krumbein, W. C.; and Pettijohn, F. J., 1938, Manual of sedimentary petrography: New York, NY, Appleton-Century-Crofts, Inc., 549 p., Chapters 5 and 6. (The settling velocities of small particles, p. 95-102; Fine sediments, p. 147-176.)

Sternberg, R. W.; and Craeger, J. S., 1961, Comparative efficiencies of size analysis by hydrometer and pipette methods: Jour. Sedimentary Petrology, v. 31, no. 1, p. 96-100.

Stokes, G. G., 1851, On the effect of the internal friction of fluids on the motion of pendulums: Trans. Cambridge Philos. Soc., v. 9, part 2, p. 8-106.

Wadell, Hakon, 1934, Some new sedimentation formulas: Physics, v. 5, p. 281-291.

__________, 1936, Some practical sedimentation formulas: Geol. Fören Forhändl., v. 58, p. 397-407.

Settling Velocity

In addition to size, other factors influence the settling rate of particles. Consider a sphere of any diameter (d), and of given density, which is settling in water. It is pulled downward by a force which depends on the particle volume, the acceleration of gravity and the difference between the density of the particle and the fluid. If the sphere is settling in water and is very small (of the order of less than 0.1 mm), the resistance of the fluid is proportional to the product of the diameter and velocity of the particle, times the viscosity of the fluid. For small spheres it can be shown that the settling velocity (v) is proportional to the square of the particle diameter, according to the equation:

$$v = Cd^2$$

where C includes the constants particle density, fluid density, acceleration of gravity and fluid viscosity. This equation is the simplest expression of Stokes' Law (1851).

Noting that most naturally occurring sedimentary particles are not spheres, Wadell (1936) derived an expression for the settling velocity of a particle intermediate in shape between a disc and a sphere. He considered this to be a better approximation of the average shape of particles in an assemblage of particles of varying shape. The data in Table 3-1 are based on Wadell's modification of Stokes' Law.

Equipment and Materials

1000 ml graduated cylinder

Beakers (50 ml)

Pipette (20 ml) with suction bulb or 10" tubing

Timer or watch with second hand

Data collection form (Form 3-1)

Distilled water

Drying oven, or hot plate, or heat lamp

Balance

Stirring rod (optional)

Dispersing agent - sodium oxalate ($Na_2C_2O_4$), or sodium hexametaphosphate ($(NaPO_4)_6$) (marketed as Calgon), or sodium lignosulfonate (Merasperse N)

Diameter (φ)	Diameter (mm)	Velocity (cm/sec)	Depth (cm)	Times of Settling (hours)	Times of Settling (minutes)	Times of Settling (seconds)
4	1/16	0.223	20	0	1	00
5	1/32	0.0558	10	0	2	59
6	1/64	0.0139	10	0	11	59
7	1/128	0.00349	10	0	47	51
8	1/256	0.00087	10	3	12	--
9	1/512	0.000217	7	8	58	--
10	1/1024	0.000054	5	25	43	--
11	1/2048	0.000013	5	106	50	--

Settling Times According to the Wadell Modification of Stokes' Law (at temperature near 20°C)

Table 3-1

Procedure

Your instructor will provide samples consisting of particles having diameters of 4φ (1/16 mm) and less. However, depending on the nature of the "fines" derived from your sieving in Exercise 2, he may ask you to use the material that you saved.

9. Continue in similar fashion, by mixing and taking remaining aliquots at specific times and depths listed in Table 3-1.

10. Dry the aliquots to constant weight in oven or on hot plate or under heat lamp at 90°C. Weigh to the nearest milligram and record weights on Form 3-1.

11. The weight of dispersent is accounted for by adding 1/50 of the total weight of dispersent to the beaker weight for each aliquot.

12. Multiply each fraction (Column 3 of Form 3-1) by 50 (if a 20 ml pipette was used), subtract successive values and record these differences as weights of phi sizes (Column 5).

13. Calculate fraction- (class-) weight percent and cumulative-weight percent.

14. You now have data which can be plotted as a cumulative-frequency curve, using the procedures that you will follow in Exercise 4. Your instructor may ask you to do this later.

$1.3\bar{3}$

$\frac{4}{3} = \quad \frac{A}{B} \quad \frac{B}{A} \quad \frac{3}{4} \quad .75$

EXERCISE 4

... if you measure that of which you speak, you know something of your subject; but if you cannot measure it, your knowledge is meager and unsatisfactory.

Lord Kelvin

EXERCISE 4
GRAPHIC PRESENTATION OF PARTICLE-SIZE DISTRIBUTION

Introduction

The distribution of particle size in sediments is a function of (1) the availability of different sizes of particles in the parent material and (2) the processes operating where the particles were deposited. In order to interpret a given sedimentary accumulation, it is necessary for the sedimentologist to present size distribution data graphically. This can be done in the form of a (1) histogram, (2) frequency curve, or (3) cumulative-frequency curve.

Graphs are constructed by plotting variables in relation to two coordinate axes at right angles to each other. The vertical axis is called the y-axis or ordinate; the horizontal axis the x-axis or abscissa. By mathematical convention, the dependent variable is plotted on the ordinate and the independent variable is plotted on the abscissa. In sediment-size distributions, particle frequency is the dependent variable and particle size is the independent variable.

The purpose of this exercise is to introduce you to techniques in (1) graphically plotting size distribution, (2) computation of statistical parameters of size distribution, and (3) interpretation of the data for purposes of inferring environment of deposition. You will use the data that you collected in "Exercise 2, Textural Analysis of Sand."

References Dealing with Graphic Presentation of Particle-Size Distribution

Dixon, W. J.; and Massey, F. J., Jr., 1957, Introduction to statistical analysis, 2nd ed.: New York, McGraw-Hill Book Company, 488 p. Chapter 1 (Distribution, p. 4-13), Chapter 3 (Measures of central value and dispersion, p. 14-29), Chapter 5 (The normal distribution, p. 48-69).

Folk, R. L.; and Ward, Wiley, 1957, Brazos River bar: a study in the significance of grain size parameters: Jour. Sedimentary Petrology, v. 27, p. 3-26.

Folk, R. L., 1966, A review of grain-size parameters: Sedimentology, v. 6, p. 73-93.

__________, 1974, Petrology of sedimentary rocks: Austin, Hemphill Publishing Co., 182 p.

Friedman, G. M., 1958, Determination of sieve-size distribution from thin-section data for sedimentary petrological (sic) studies: Jour. Geology, v. 66, p. 394-416.

__________, 1961, Distinction between (sic) dune, beach and river sands from their textural characteristics: Jour. Sedimentary Petrology, v. 31, p. 514-529.

__________, 1962, On sorting, sorting coefficients and the lognormality of the grain-size distribution of sandstones: Jour. Geology, v. 70, p. 737-753.

__________, 1967, Dynamic processes and statistical parameters compared for size frequency distribution of beach and river sands: Jour. Sedimentary Petrology, v. 37, p. 327-354.

__________, 1979, Differences in size distributions of populations of particles among sands of various origins: Sedimentology, v. 26, p. 3-32.

__________, 1979, Differences in size distributions of populations of particles among sands of various origins: Addendum to IAS presidential address: Sedimentology, v. 26, p. 859-862.

__________; and Sanders, J. E., 1978, Principles of sedimentology: New York, NY, John Wiley & Sons, 792 p., Chapter 3 (Size distributions of populations of particles in sedimentary deposits, p. 70-81).

Griffiths, J. C., 1967, Scientific method in analysis of sediments: New York, McGraw-Hill Book Co., 508 p., Chapter 6 (Grain shape, p. 109-143), Chapter 7 (Fabric, p. 144-163).

Inman, D. C., 1952, Measures for describing the size distribution of sediments: Jour. Sedimentary Petrology, v. 22, p. 125-145.

Krumbein, W. C., 1933, Textural and lithological variations in glacial (sic) till: Jour. Geology, v. 41, p. 382-408.

__________, 1936, The application of logarithmic moments to size frequency distributions of sediments: Jour. Sedimentary Petrology, v. 6, p. 35-47.

__________; and Pettijohn, F. J., 1938, Manual of sedimentary petrography: New York, NY, Appleton-Century-Crofts, Inc., 549 p., Chapter 7 (Graphic presentation of analytical data, p. 182-211); Chapter 8

(Elements of statistical analysis, p. 212-227); Chapter 9 (Application of statistical measures to sediments, p. 228-267).

Material

Data from Form 2-1 of "Exercise 2, Textural Analysis of Sand."

The Normal Distribution

A frequency-distribution curve is a plot of the frequency with which some variable occurs within an arbitrarily defined subclass of a population. In sedimentology, a size-frequency curve is made by plotting the proportion, in percent, of various particle sizes, the class limits of which are expressed in phi units. By convention, frequency is plotted along the ordinate and phi values, increasing to the right, along the abscissa. If particle size is distributed normally, the central part of the population contains the bulk of the particle sizes and the coarser and finer fractions are distributed on each side of the central part of the curve (Fig. 4-1).

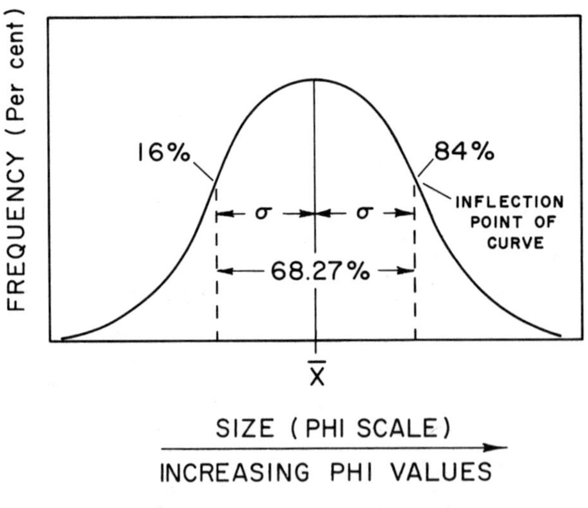

Normal Frequency-Distribution Curve
(Friedman and Sanders, 1978, Fig. 3-12, p. 70)

Figure 4-1

To put this more directly, by definition, all normal distributions plot as symmetrical, bell-shaped curves and, in addition, must satisfy other requirements: (1) the two points of inflection on the flanks of the curve are located such that the area under the curve between these points is equal to 68.27% of the area under the entire curve; and (2) half of this central 68.27%, or 34.13%, lies on each side of the center of the distribution (the 50% point). The borders of the central 68.27% of the curve lie at percentage values of 16 and 84%. These departures of 34.13% to the left and 34.13% right of the middle of the distribution are referred to as the standard deviation (σ). The highest point on the curve, which indicates the most abundant class in the population, is the mode. The midpoint of the distribution, the point that divides the area under the curve into two equal parts, is the mean ($\bar{x}$). For a normal distribution, the mode and mean are the same value.

Most distribution curves for sediment samples are not normal in the sense that they lack the symmetry of the curve shown in Figure 4-1. Later in this exercise you will plot a frequency-distribution curve for data on Form 2-1 and use curve asymmetry as an aid in interpreting processes of sediment accumulation.

The Histogram

A histogram is an easily prepared graphic representation of frequency-distribution data. It consists of bars arranged vertically along a horizontal scale representing various class limits. Figure 4-2 is a histogram showing the same particle-size data that are shown in the normal frequency-distribution curve of Figure 4-1.

Compare Figures 4-1 and 4-2. Although the general shapes of the two plots are the same, you should be aware that there are important differences between them. First, the histogram cannot be used for determination of statistical parameters; second, its shape is greatly affected by the sieve interval which is chosen for the sieving analysis. Consequently, although the histogram is useful for depicting the general textural features of a sedimentary deposit, it is useless for detailed analysis.

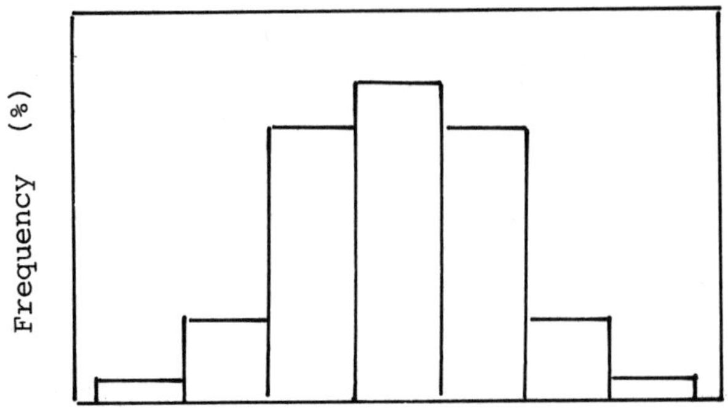

Histogram of Particle-Size Data shown in Figure 4-1.

Figure 4-2

Procedure I: Plotting Data as a Histogram

Using data from Form 2-1, plot a histogram for your sand-size analysis on Form 4-1. Does the histogram indicate that your sample has a normal size-frequency distribution? If not, in what way does it differ from a normal distribution?

Variations in Shape of the Frequency Curve

Curves of normal distributions having the same means and enclosing the same total area may have different shapes (Fig. 4-3). This difference in shape indicates that the number of phi classes on the abscissa between the points of inflection will not be the same for each curve. If a curve is steep (curve A in Fig. 4-3), the number of phi classes between the two points is small; if the curve is flat and wide (curve C in Fig. 4-3), the number of phi units between the points is large. This spread of the distribution with respect to the mean defines the sorting of the material. Well-sorted sediment has a sharp-peaked, narrow curve. Poorly sorted sediment has a low, wide curve.

The three curves shown in Figure 4-3 all show a normal distribution of particle size; that is, they are symmetrical and satisfy the requirements outlined following Figure 4-1. However, some frequency-distribution curves exhibit a characteristic called skewness (Fig. 4-4).

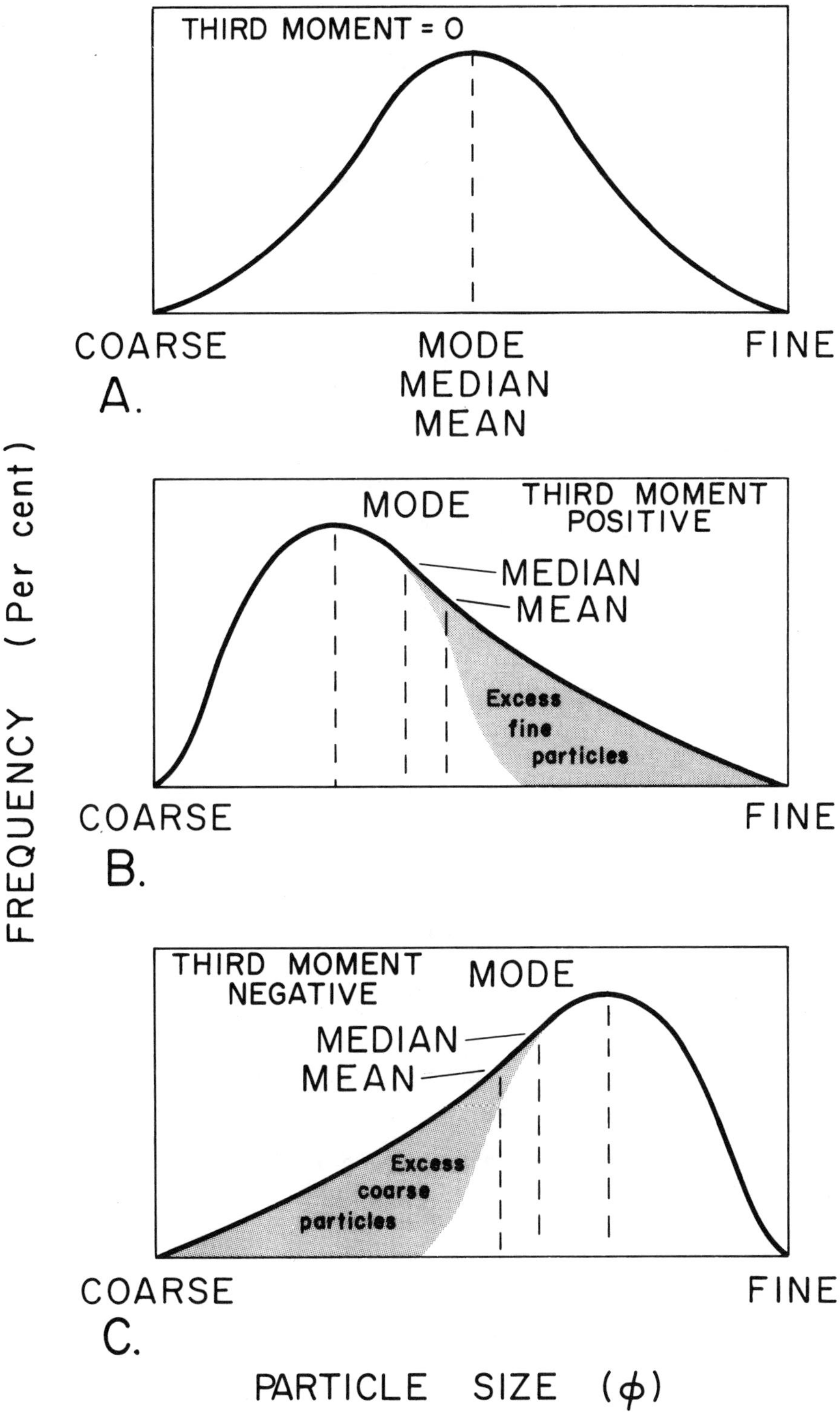

Frequency Distribution Curves having Various Skewnesses
(Friedman and Sanders, 1978, Fig. 3-18, p.75)

Figure 4-4

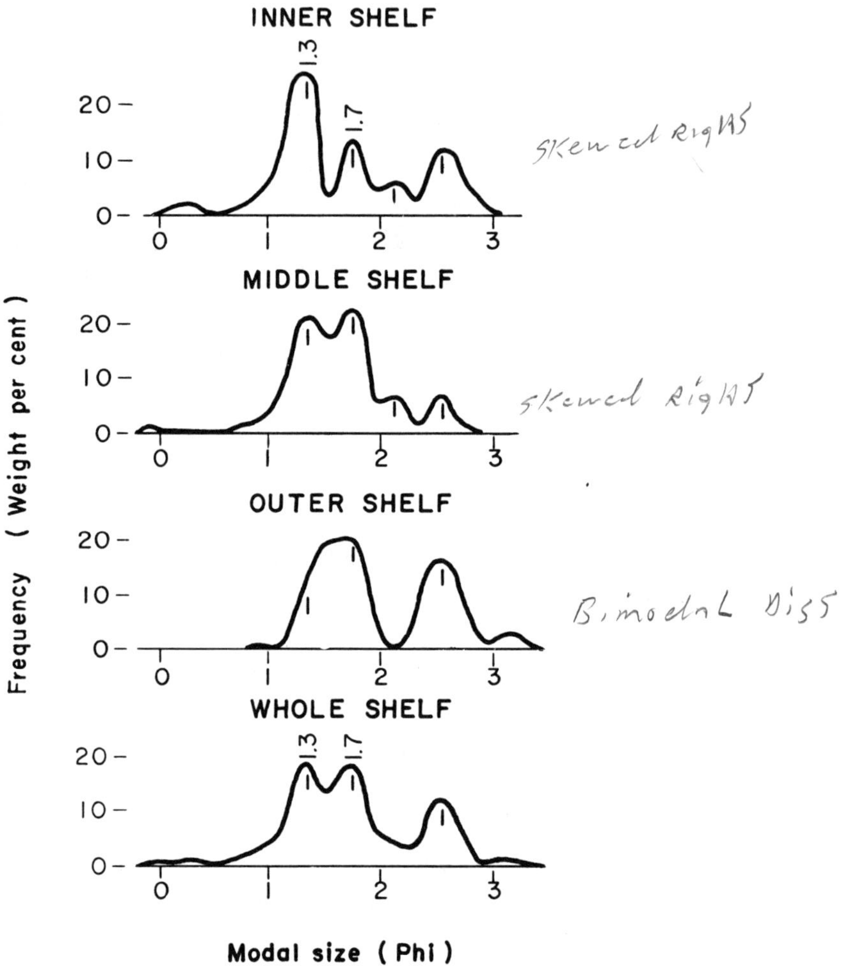

Frequency distributions of sands from continental shelf off Long Island, New York, showing various modes. Interpretations of McKinney and Friedman (1970, Fig. 17, p. 233).

Figure 4-5

FREQUENCY-DISTRIBUTION CURVE - Results of Exercise 2 Sieve Analysis

Exercise 4
Name ____________________
Laboratory Section ____________________
Date ____________________

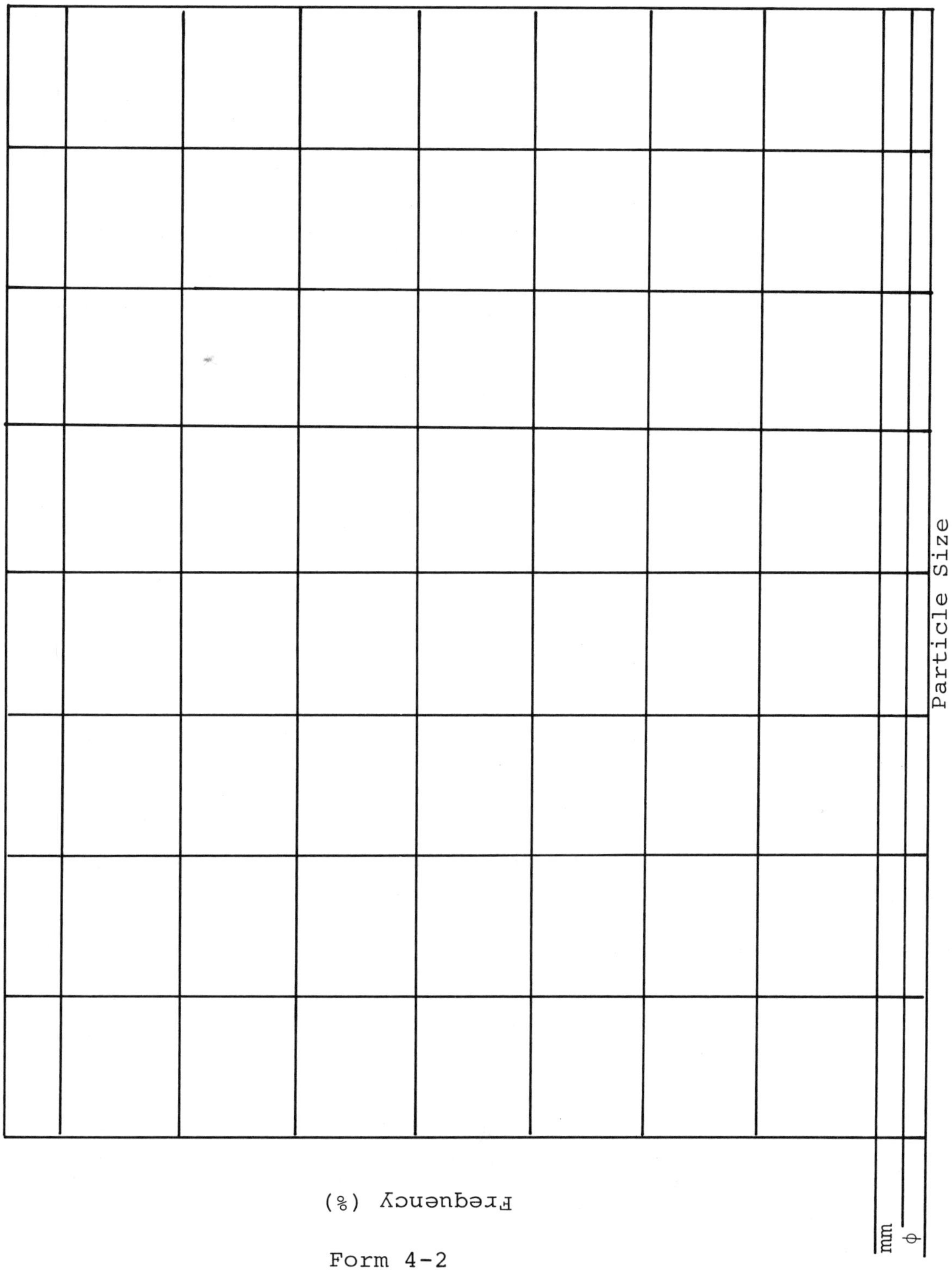

Form 4-2

The Cumulative-Frequency Curve

In studies of populations of sedimentary particle sizes, another kind of plot, the cumulative-frequency curve, is particularly valuable. The cumulative-frequency curve differs from the simple frequency curve in that each point represents not just the frequency of each class but rather, the sum of all percentages of the preceding (coarser) size classes. A cumulative-frequency curve starts at zero percent at the lower left corner of the graph and climbs upward and towards the right toward 100 percent as each frequency class is added. Various kinds of ordinate scales are used, the most common being arithmetic or probability. The abscissa scale consists of phi units increasing towards the right.

When normal distributions are plotted using an arithmetic ordinate scale, S-shaped, or ogive curves result (Fig. 4-6).

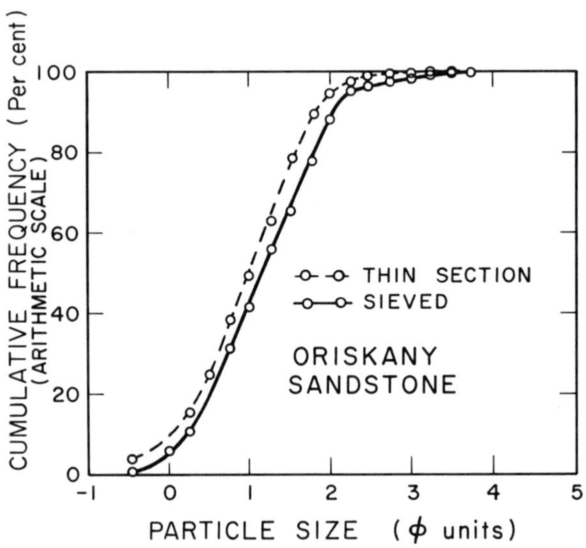

Cumulative-Frequency Curve having an Arithmetic Ordinate Scale
(Friedman, 1958, in Friedman and Sanders, 1978, Fig. 3-14A, p. 72)

Figure 4-6

Procedure III: Plotting Data as a Cumulative-Frequency Curve (Arithmetic Ordinate)

Using data from Form 2-1, plot a cumulative-frequency curve on Form 4-3. The advantage of this curve is that all statistical parameters may be read from it exactly and that the curve is independent of the sieves used.

When cumulative-frequency distributions of the particle sizes of sediments are plotted on probability graphs, one of the most striking observations is that most of them do not come out as one continuous straight line. In fact, commonly two or three or even more straight-line segments are present, each having a different slope and separated by a sharp break between the segments (Fig. 4-8). These segmented curves indicate the

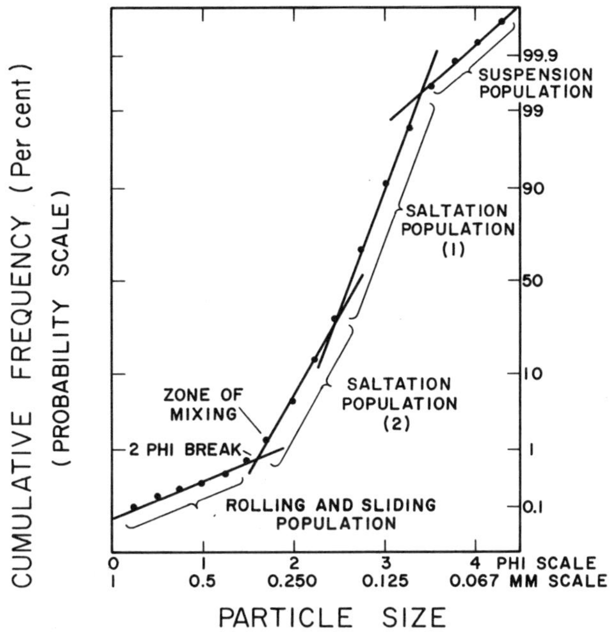

Cumulative-frequency curve of beach sand, drawn on probability paper, consisting of four straight-line segments. (G. S. Visher, 1969, Fig. 4, p. 1079, in Friedman and Sanders, 1978, Fig. 3-15, p. 72).

Figure 4-8

inappropriateness of considering that the entire sample constitutes one population having a single normal distribution. The slope of each straight-line segment and the position of the breaks between segments reflect the mechanism of deposition. In sands with three straight-line segments, the central and larger segment may represent a subpopulation that moved in the current by a jumping motion, saltation, the segment at the fine end of the distribution resulted from deposition of particles carried in suspension, and that at the coarse end from particles deposited by rolling or sliding (Fig. 4-8).

Procedure IV: Plotting Data as a Cumulative-Frequency Curve (Probability Ordinate)

Using data from Form 2-1, plot a cumulative-frequency curve on Form 4-4. Compare the curve plotted on an arithmetic ordinate (Form 4-3) with this one, plotted on a probability ordinate. Summarize the advantages and disadvantages of the two forms of graphic presentation of particle-size data in the space below:

CUMULATIVE-FREQUENCY CURVE
(PROBABILITY ORDINATE) - Results
of Exercise 2 Sieve Analysis

Exercise 4
Name ______________________
Laboratory Section ______________
Date ______________________

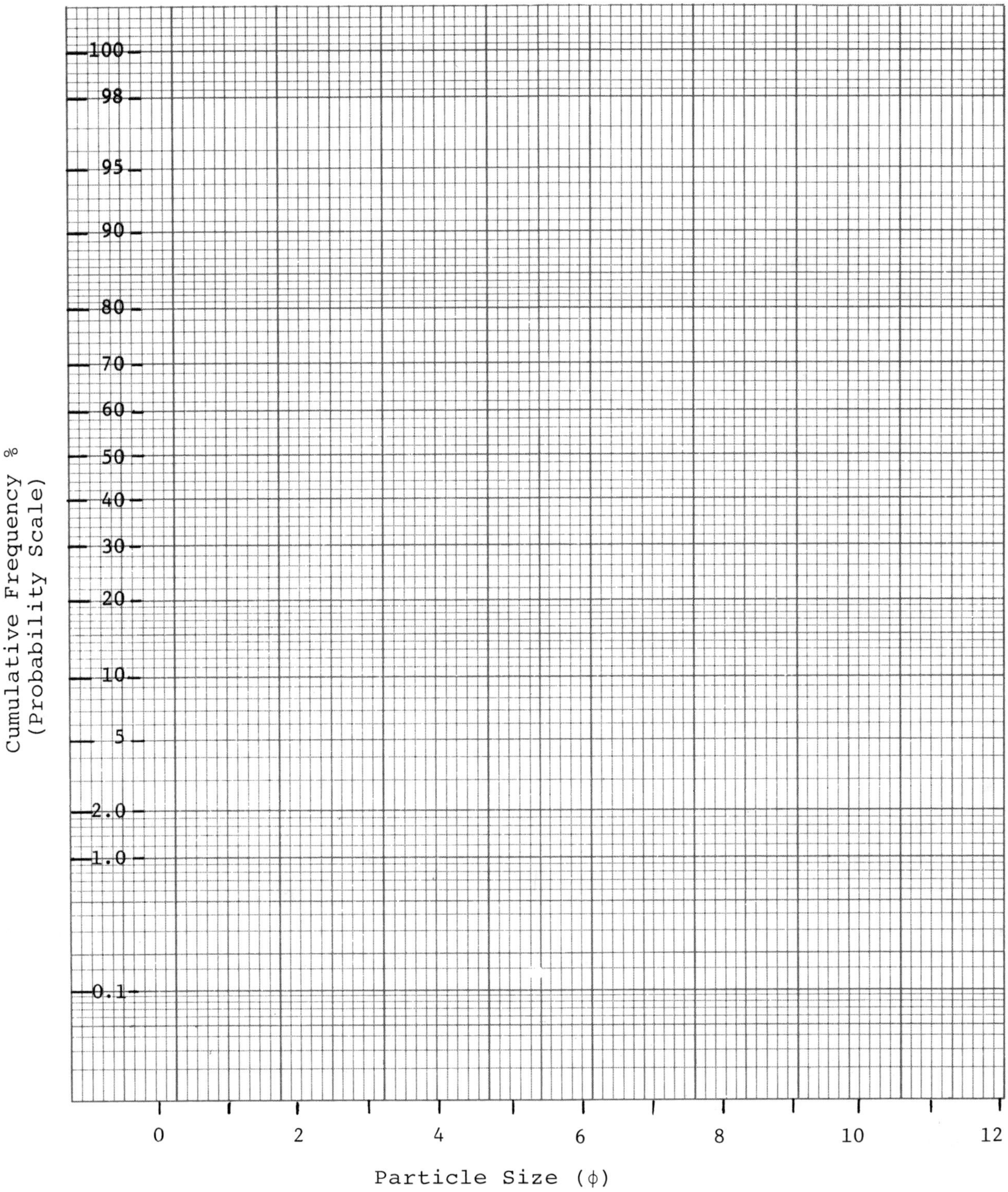

Form 4-4

Graphic Measures of the Frequency Distribution

The statistical measures of a frequency distribution can be determined graphically. In this method one uses the cumulative-frequency curve to determine the phi values of various percentile points. The values are then inserted into formulas designed to approximate the values computed by using moments. (In conventional statistics, the normal distribution is defined in terms of moments, a term introduced into statistics by analogy with mechanics.) The graphic approach was devised at a time when it was necessary to compute moments by tedious hand calculation. Although it is now possible to routinely compute moments with modern, high-speed computers, as well as simple hand calculators having memory stores, the graphic approach is still widely used by sedimentologists to generate data useful for comparing suites of samples or for interpreting sedimentary processes.

In the graphic approach, phi values from the curve are read for various percentiles. Many investigators have employed the 25th and 75th percentiles, which define the so-called quartile deviations. The quartile deviations represent the central part of the curve and hence involve the bulk of the particles. Other workers rely on the 16th and 84th percentiles, which represent sizes lying one standard deviation on either side of the mean in a normal distribution. These values may be combined with those of the 5th and 95th percentiles, which are convenient numbers for expressing the ends of the distribution.

The median is the particle size in the exact middle of the population; half the particles (by weight) are finer and half coarser. The median can be read directly from the cumulative curve; it is the phi intercept of the point where the cumulative-frequency curve crosses the 50 percent mark on the cumulative-frequency scale. Another size factor that can be read from a graph is the mode, the most-frequently occurring particle size. The mode is the peak of a simple frequency curve. It cannot generally be determined from the cumulative-frequency curve. As indicated on page 55, in a normal distribution, the phi values of mode, median and mean coincide; in other distributions, they do not.

Once the various percentile values have been read from the probability plot, the approximate statistical parameters are determined using the following formulas (the subscripts are the percentile values read from a probability plot of the cumulative-frequency curve):

Graphic mean $$(M_z) = \frac{\phi_{16} + \phi_{50} + \phi_{84}}{3}$$

Inclusive graphic standard deviation $$(\sigma_1) = \frac{\phi_{84} - \phi_{16}}{4} + \frac{\phi_{95} - \phi_5}{6.6}$$

Simple sorting measure $$(So_s) = \tfrac{1}{2}(\phi_{95} - \phi_5)$$

Inclusive graphic skewness $$(SK_1) = \frac{(\phi_{84} + \phi_{16} - 2\phi_{50})}{2(\phi_{84} - \phi_{16})} + \frac{(\phi_{95} + \phi_5 - 2\phi_{50})}{2(\phi_{95} - \phi_5)}$$

Simple skewness measure $$(\alpha_s) = (\phi_{95} + \phi_5) - 2\phi_{50}$$

Graphic kurtosis $$(K_G) = \frac{(\phi_{95} - \phi_5)}{2.44(\phi_{75} - \phi_{25})}$$

Procedure V: Computation of Approximate Statistical Parameters of the Particle-Size Distribution

Using data taken from Form 4-4, compute the (1) graphic mean, (2) inclusive graphic standard deviation, (3) simple sorting measure, (4) inclusive graphic skewness, (5) simple skewness measure, and (6) graphic kurtosis. Record these parameters on Form 4-5.

Interpretation of Data

The size distribution of particles in a sand will be influenced greatly by the conditions that determine what happens to the fine particles forming the suspended load. Sands deposited by rivers almost invariably contain fine particles from the suspended load. Sands deposited on the parts of beaches where the breaking waves continuously wash thin sheets of water back and forth invariably lack admixtures of fine-grained sediments. Water motion is so vigorous that the fine particles are always kept moving and do not come to rest with the sands. For these kinds of sands the positions of the breaks between the straight-line segments on the probability plots are predictable and are functions of the processes that deposited the sediment.

These relationships between the contrasting unidirectional flow of water in rivers and bidirectional flow of water on parts of beaches and

STATISTICAL PARAMETERS COMPUTED
FROM CUMULATIVE-FREQUENCY CURVE
Results of Exercise 2 Sieve
Analysis

Exercise 4
Name ____________________
Laboratory Section ____________________
Date ____________________

Parameter	Formula	Value
Graphic mean	$(M_z) = \dfrac{\phi_{16} + \phi_{50} + \phi_{84}}{3}$	
Inclusive graphic standard deviation	$(\sigma_1) = \dfrac{\phi_{84} - \phi_{16}}{4} + \dfrac{\phi_{95} - \phi_5}{6.6}$	
Simple sorting measure	$(So_s) = \frac{1}{2}(\phi_{95} - \phi_5)$	
Inclusive graphic skewness	$(SK_1) = \dfrac{(\phi_{84} + \phi_{16} - 2\phi_{50})}{2(\phi_{84} - \phi_{16})} + \dfrac{(\phi_{95} + \phi_5 - 2\phi_{50})}{2(\phi_{95} - \phi_5)}$	
Simple skewness measure	$(\alpha_s) = (\phi_{95} + \phi_5) - 2\phi_{50}$	
Graphic kurtosis	$(K_G) = \dfrac{(\phi_{95} - \phi_5)}{2.44(\phi_{75} - \phi_{25})}$	

their characteristic size distributions, so clearly indicated by the cumulative-probability plots, likewise become apparent from applications of the older assumption that only one statistical size function covers all size classes of a sample. The following paragraphs provide a comparison of the probability-plot approach and the one-function (frequency curve) approach to interpretation of particle size data.

Probability plots of river sands typically show two straight-line segments with a break commonly occurring at a cumulative frequency of less than 90 percent and falling within the range between 2.5 and 3.5ϕ (Fig. 4-9A). Thus, the segment representing the fine fraction usually constitutes 10 percent or more of the total distribution. The fraction has resulted from the trapping of suspended particles among coarser particles during movement of the current and from deposition as the current slackened.

On the frequency curve using the "one-function" approach, the presence of these abundant fine particles is shown by the stretching out or "trailing off" of the curve to the right of the mean (Fig. 4-4B). Such a frequency distribution is positively skewed. Despite many exceptions, the frequency distributions of most river sands display positive numerical values for skewness. This results in large part from the presence of the fine-grained fraction (Fig. 4-4B).

Probability plots of one variety of beach sand, the kind deposited by the bidirectional water motions mentioned previously, indicate that the segment representing the suspended particles usually constitutes less than one percent (Fig. 4-9B). In addition, the central parts of the graphs tend to show two separate straight-line segments, each representing a different saltation population.

The one-function approach shows that frequency distributions of such beach sands tend to be symmetrical (Fig. 4-4A), negatively skewed (Fig. 4-4C), or slightly positively skewed. These frequency distributions are not stretched out to the right of the mean, as were the distribution curves of the river sands.

As mentioned previously, the slope of the cumulative curve on a probability plot reflects the degree of sorting. This is true as well for the general trend of the entire sample even if several subpopulations are present as it is for each straight-line segment. Hence, the standard deviation must be a process-sensitive parameter, even if we apply the one-function approach. Comparison of Figure 4-9A with Figure 4-9B shows how

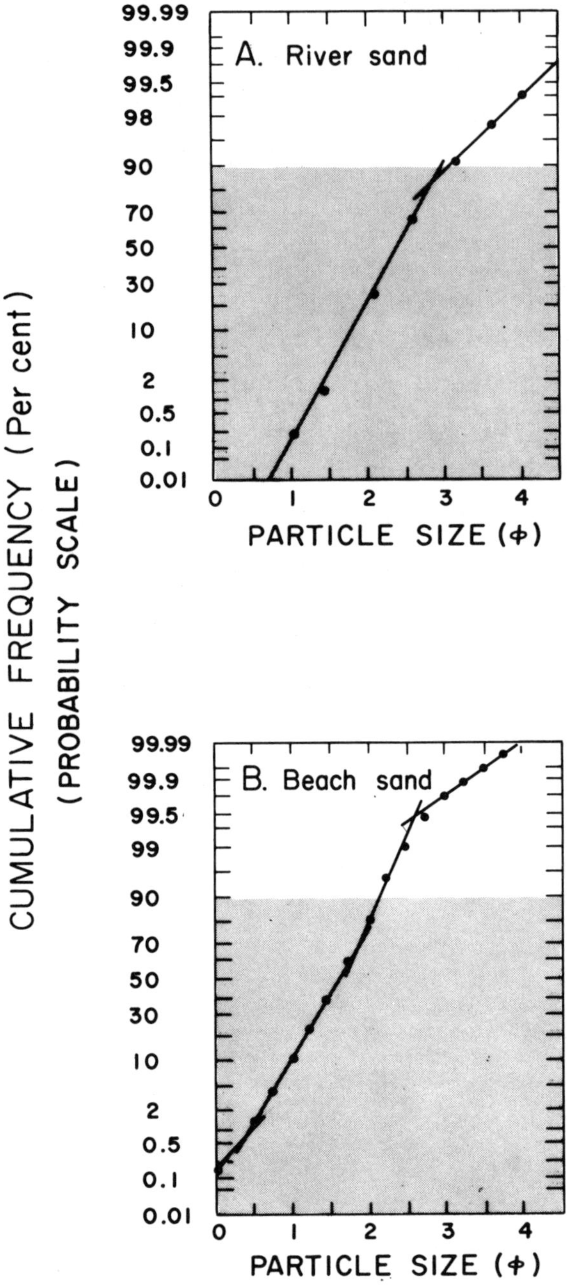

Cumulative-frequency curves: A. River sand (Arkansas River);
B. Beach sand (Bald Point, Florida).
(G. S. Visher, 1965, Fig. 10, p. 126, and 1969, Fig. 7, p. 1082, in Friedman and Sanders, 1978, Fig. 3-20, p. 77.)

Figure 4-9

slopes of each straight-line segment compare with the general slope of the entire curve (the line that would be formed by connecting the lowermost point at the left with the uppermost point at the right). The slopes of the segments representing the saltation populations are generally similar for both river and beach sands. Although the slope of the entire distribution for the beach sand in Figure 4-9B is parallel to the slope of the entire distribution of the stream sand in Figure 4-9A, in many other examples beach sands are better sorted than river sands and therefore the slopes of curves for beach sands tend to be steeper than those of river sands. This same relationship emerges by a computation of the standard deviation for beach and river sands using the one-function approach. Because of the effects of the bidirectional flow of thin sheets of water from the breaking waves, beach sands subjected to such oscillating flow are generally well sorted; the numerical values for their standard deviations are low. By contrast, river sands generally tend to be less well sorted than such beach sands. Hence the numerical values for the standard deviation of river sands are higher. As noted previously, the skewness is sensitive to geologic processes. Therefore a scatter plot of the two process-sensitive factors, standard deviation (sorting) against skewness, is likely to be significant. On such a plot, despite some overlap, the beach sands (of the kind we have been discussing) occupy one area and the river sands another (Fig. 4-10). Whereas the spread of values for sorting

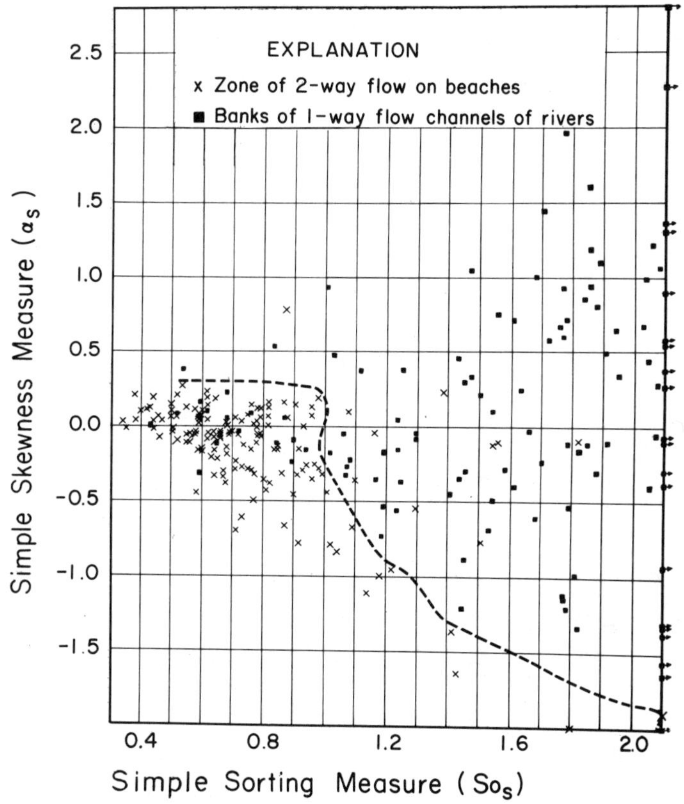

Scatter plot, simple skewness measure versus simple sorting measure, for beach and river sands. (G. M. Friedman, 1967, Fig. 18, p. 342, in Friedman and Sanders, 1978, Fig. 3-21, p. 78.)

Figure 4-10

and skewness indicates that the frequency distribution for river sands varies considerably, the spread of values for beach sands is much less variable. A scatter plot, as shown in Figure 4-10, just like the straight-line segments of probability plots, reveals how a study of sizes of particles relates to processes forming sedimentary deposits.

Similarly, plots have been made (Friedmann, 1979) which show that sorting can be used to discriminate between (1) beach and inland-dune sands (Figure 4-11), and (2) between inland and nearshore dune sands (Figure 4-12).

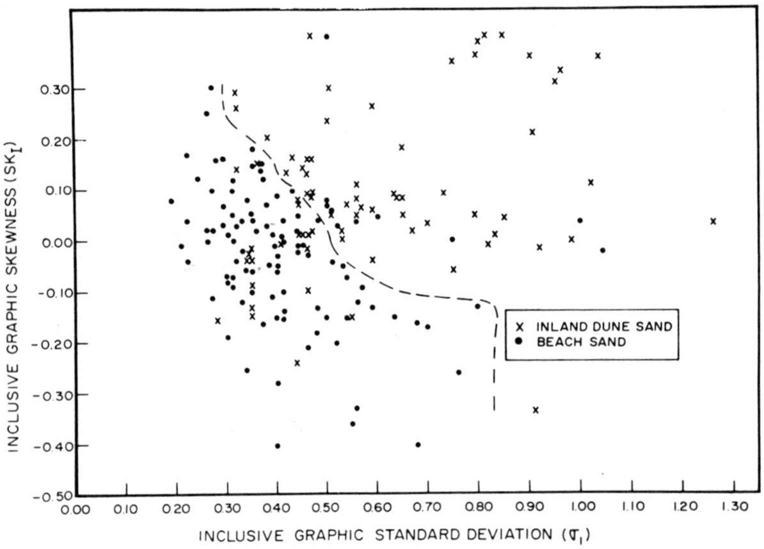

Plot of inclusive graphic standard deviation and inclusive graphic skewness for beach and inland-dune sands. (Friedman, 1979, p. 860.)

Figure 4-11

Although quantitative studies of size and sorting and various statistical parameters have become a part of modern sedimentological practice, it is worth noting that one need not depend entirely on them. Observations with the hand lens can go a long way in relating size fractions to depositional processes. If silt and clay are present in a sand, then that sand probably was not deposited in the zone of bidirectional water motion on a beach where the fine fractions remain suspended and are deposited somewhere else. Hence, such beach sands, and coastal dune sands derived from them, tend to lack silt and clay.

Procedure VI: Interpretation of Particle-Size Distribution Data

Based on the size-distribution data in Forms 4-1, 4-2, 4-3, 4-4, and 4-5 discuss in the space below the probable physical conditions under which the sediment was deposited.

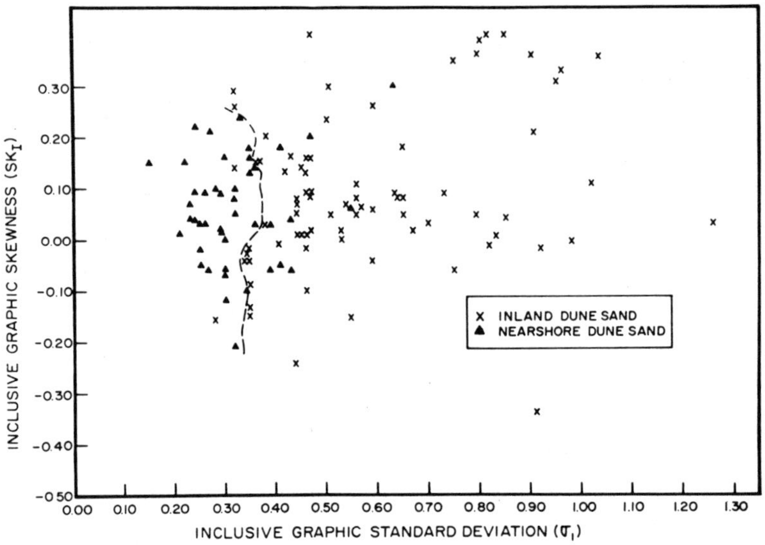

Plot of inclusive graphic standard deviation and inclusive graphic skewness for inland and nearshore dune sands. (Friedman, 1979, p. 860.)

Figure 4-12

Moment Measures

Depending on the specific type of sedimentology course that you are taking or your objectives in learning about sedimentology, you may be interested in knowing more about moment measures. In conventional statistics, the normal distribution is defined in terms of moments. The term moment was introduced into statistics by analogy with mechanics. In mechanics, the moment of a force about a point of rotation, such as a fulcrum, is computed by multiplying the magnitude of that force times the distance from the force to the point of rotation (Fig. 4-13). In statistical moments the force of mechanics is replaced by a frequency function (such as the percent of the distribution within the given class interval or, when related to the sizes of the particles, the percent of the grains falling within a certain size class). The concept of the distance to a point is the same in both kinds of moments. In statistical moments, the point of rotation of mechanics is replaced by an arbitrary point, usually either the origin of the curve or its mean. The distance is determined by the spacing of the size class (shaded rectangle in Fig. 4-13) from the arbitrary point.

This statistical moment, the frequency percent times distance (Fig. 4-13) as just described, is the moment of that size class with

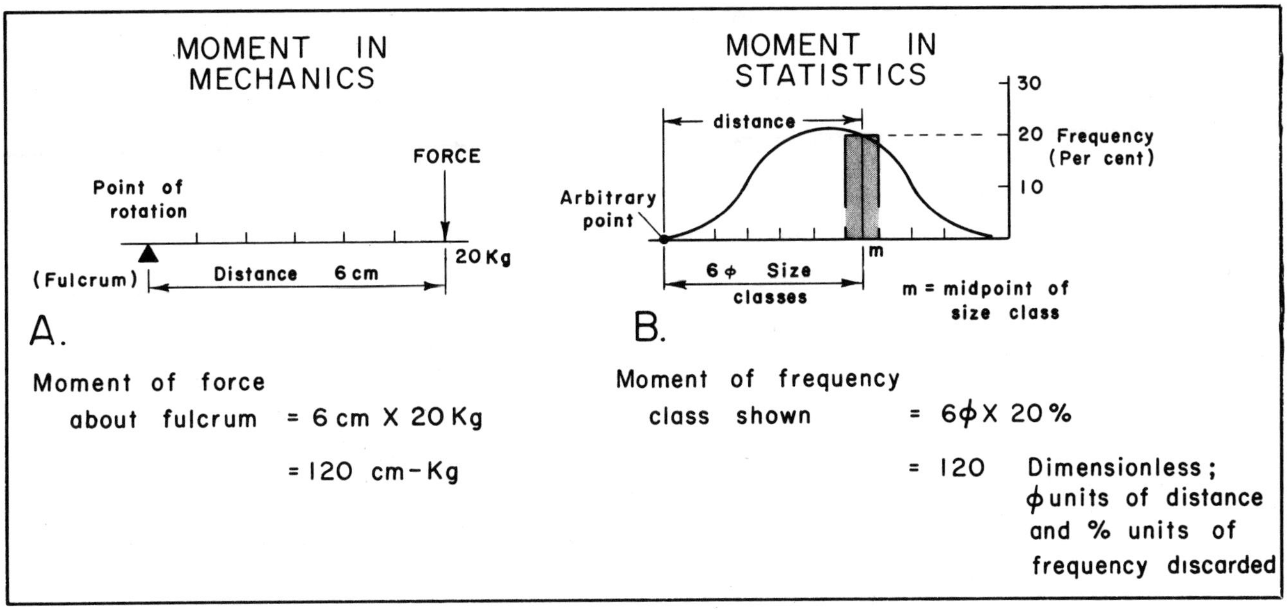

Moments in mechanics and statistics.
(Friedman and Sanders, 1978, Fig. 3-22, p. 79.)

Figure 4-13

respect to the arbitrary point (i.e., to the origin of the curve). The statistical moment of the distribution is the moment per unit frequency. It is determined by finding the moments of each size class (frequency percent in each size class times the distance of each from the origin, as just described), adding them up, and dividing the sum by 100. This gives the moment per unit (1 percent) frequency. In mathematical notation the first moment of the distribution is written as follows:

$$\text{first moment} = \frac{\sum fm_\phi}{100}$$

where f is frequency in percent for each size class, and m_ϕ is the midpoint of each ϕ size class. By definition, the first moment equals the mean, $\bar{x}$, therefore

$$\bar{x} = \frac{\sum fm_\phi}{100}$$

Having determined the mean, we can, if we like, take another first moment by using the mean itself as the arbitrary point. As we do so, the values for our distance function will be positive to the right of the mean and negative to the left of it. The distribution is normal if the algebraic sum of the positive and negative numbers is zero. Notice that in the first moment, the distance term (which is m_ϕ) appears in the first power. Successively higher moments are defined by raising the distance term to progressively higher powers. Thus, the distance term is squared in the second moment, cubed in the third moment, raised to the fourth power in the fourth moment, and to the nth power in the nth moment. After the mean has been defined, it is convenient to use it as the arbitrary point for the distance function in the higher moments. Therefore, in these higher moments, the distance term will be represented with respect to the mean, $\bar{x}$, by $(m_\phi - \bar{x})$ (Fig. 4-14). The higher moments contrast with the first moment in that they are moments about the mean, whereas the first moment is the moment with respect to an arbitrary point, such as the origin of the curve.

The second moment of the distribution is a measure of dispersion about the mean and is expressed mathematically as

$$\text{second moment} = \frac{\sum f(m_\phi - \bar{x})^2}{100}$$

This second moment represents the numerical value of the standard deviation squared. In order to obtain the numerical value of the standard deviation, σ, we take the square root of the second moment, as follows:

$$\sigma = \left[\frac{\sum f(m_\phi - \bar{x})^2}{100}\right]^{1/2}$$

As already noted, in populations of particles, the standard deviation provides information on the extent to which particle sizes are clustered about the mean, and hence defines the concept of sorting.

The third moment of the distribution is a measure of the symmetry of the frequency curve about the mean and is written as follows:

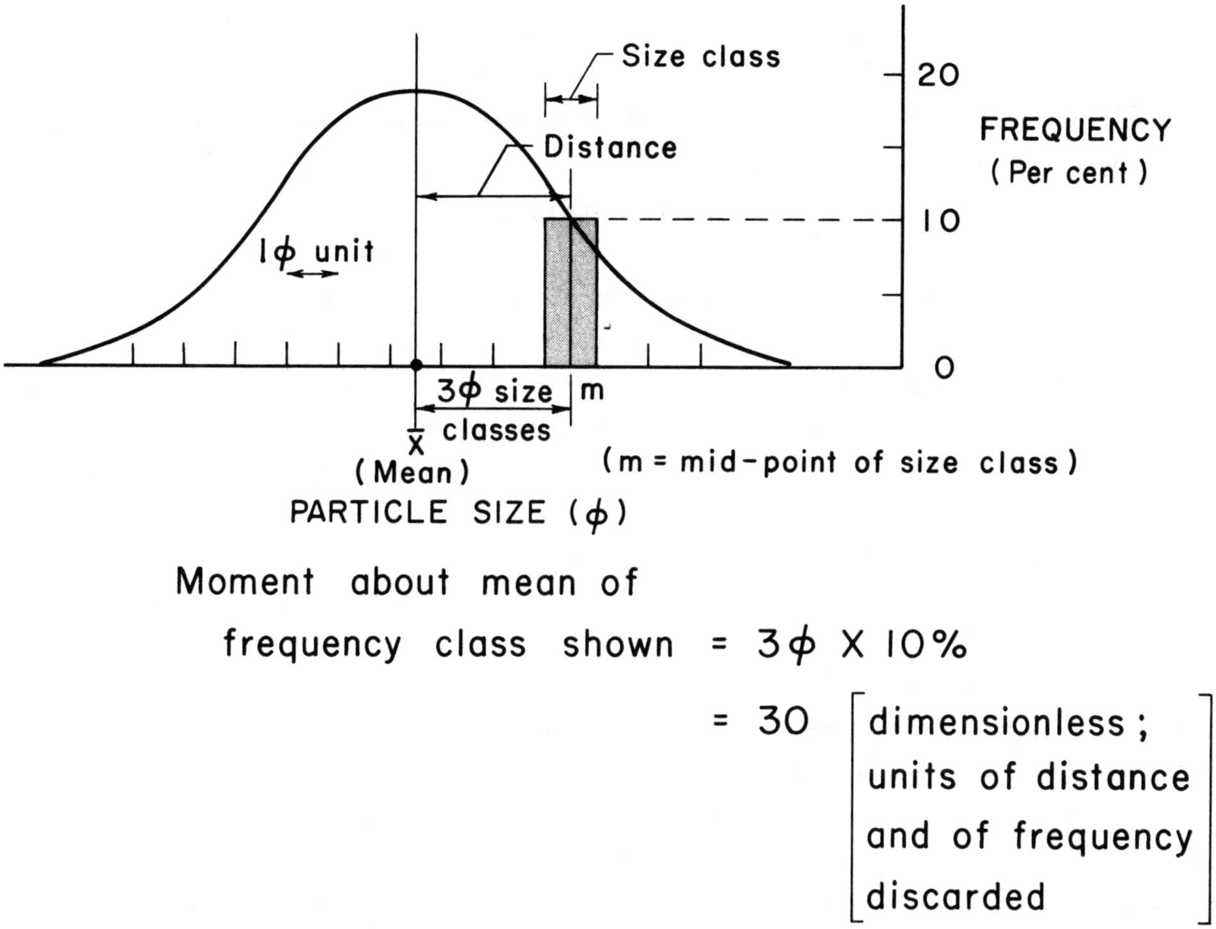

Moment about mean.
(Friedman and Sanders, 1978, Fig. 3-23, p. 79.)

Figure 4-14

$$\text{third moment} = \frac{\sum f(m_\phi - \bar{x})^3}{100}$$

This moment is known as the mean cubed deviation and, by testing the symmetry of the curve, determines its normality. As $(m_\phi - \bar{x})$ is positive to the right of the mean and negative to the left of the mean, the value for third moment in a normal distribution is 0 (Fig. 4-4A). As already noted, a positive value indicates an excess of particles (shown by a trailing off of the curve) in that part of the distribution lying to the right of the mean (positively skewed distribution, Fig. 4-4B). By contrast, a negative number means an excess of particles to the left of the mean (negatively skewed distribution, Fig. 4-4C). In other words, the third moment yields deviation from symmetry, a concept known as skewness. This measure is important in the study of particle sizes as it indicates that particles in excess of the normal distribution are present in the coarser fraction or finer fraction, especially in the extremes (the so-called tails) of the distribution. This information can be used in interpreting how particles were deposited.

The skewness of the curve commonly is derived by dividing the mean cubed deviation of the cube of the standard deviation. This is expressed by the equation:

$$\text{skewness}(\alpha_3) = 1/100\ \sigma^{-3} \sum f(m_\phi - \bar{x})^3$$

Skewness, as the third moment, reflects deviation from symmetry of the curve and is sensitive to the presence or absence of the fine fraction and coarse fraction in populations of particles and hence has been used in interpreting the deposition of sediments.

The fourth moment of the distribution about the mean is expressed mathematically as

$$\text{fourth moment} = \frac{\sum f(m_\phi - \bar{x})^4}{100}$$

The fourth moment is used to calculate the peakedness, or kurtosis of the distribution. Kurtosis is determined by dividing the fourth moment by the standard deviation raised to the fourth power, hence:

$$\text{kurtosis}(\alpha_4) = 1/100\ \sigma^{-4} \sum f(m_\phi - \bar{x})^4$$

A curve of a frequency distribution may satisfy the requirements of the first three moments, but not the fourth moment. Specifically, this means that the curve may be symmetrical ($\alpha_3 = 0$) and the middle 68 percent of the distribution lies between the 16- and 85-percent points (second moment or standard deviation), yet may have a sharper (leptokurtic, Fig. 4-15) or a flatter (platykurtic, Fig. 4-15) peak than the normal (mesokurtic) curve. According to the equation for the normal distribution, given below, a normal curve must have a kurtosis value of 3.

The equation of the normal distribution is written

$$Y = \frac{1}{\sigma(2\pi)^{\frac{1}{2}}}\, e^{\frac{-\frac{1}{2}(m - \bar{x})^2}{\sigma}}$$

where π is a constant of approximately 3.1416 and e is a constant of approximately 2.7183.

For most practical purposes, a frequency distribution may be considered to be normal if its skewness is 0 and its kurtosis is 3. Theoretically, a normal curve has to satisfy higher moments in addition to the third moment and fourth moment.

Recalculating the data for beach sands and river sands (Fig. 4-10) by substituting moment measures for graphic measures and plotting the two process-sensitive factors standard deviation against the third moment (mean cubed deviation) gives a better separation between the fields that beach sands and river sands occupy than graphic measures achieved (Fig. 4-16). Moment measures of the size distribution are more sensitive to environmental processes than graphic measures.

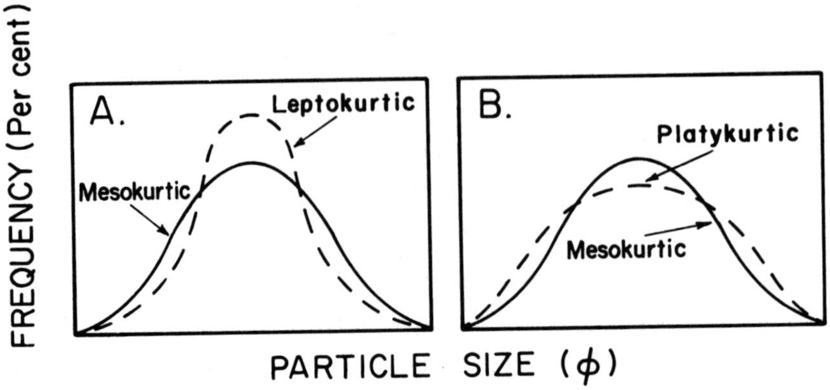

Frequency distribution curves having various kurtoses.
(Friedman and Sanders, 1978, Fig. 3-24, p. 80.)

Figure 4-15

References Dealing With Moment Measures

Dixon, W. J., and Massey, F. J., Jr., 1957, Introduction to statistical analysis, 2nd ed.: New York, McGraw-Hill Book Company, 488 p., Chapter 1 (Distributions, p. 4-13), Chapter 3 (Measures of central value and dispersion, p. 14-29), Chapter 5 (The normal distribution, p. 48-69).

Griffiths, I. C., 1967, Scientific method in analysis of sediments: New York, McGraw-Hill Book Co., 508 p., Chapter 6 (Grain shape, p. 109-143), Chapter 7 (Fabric, p. 144-168).

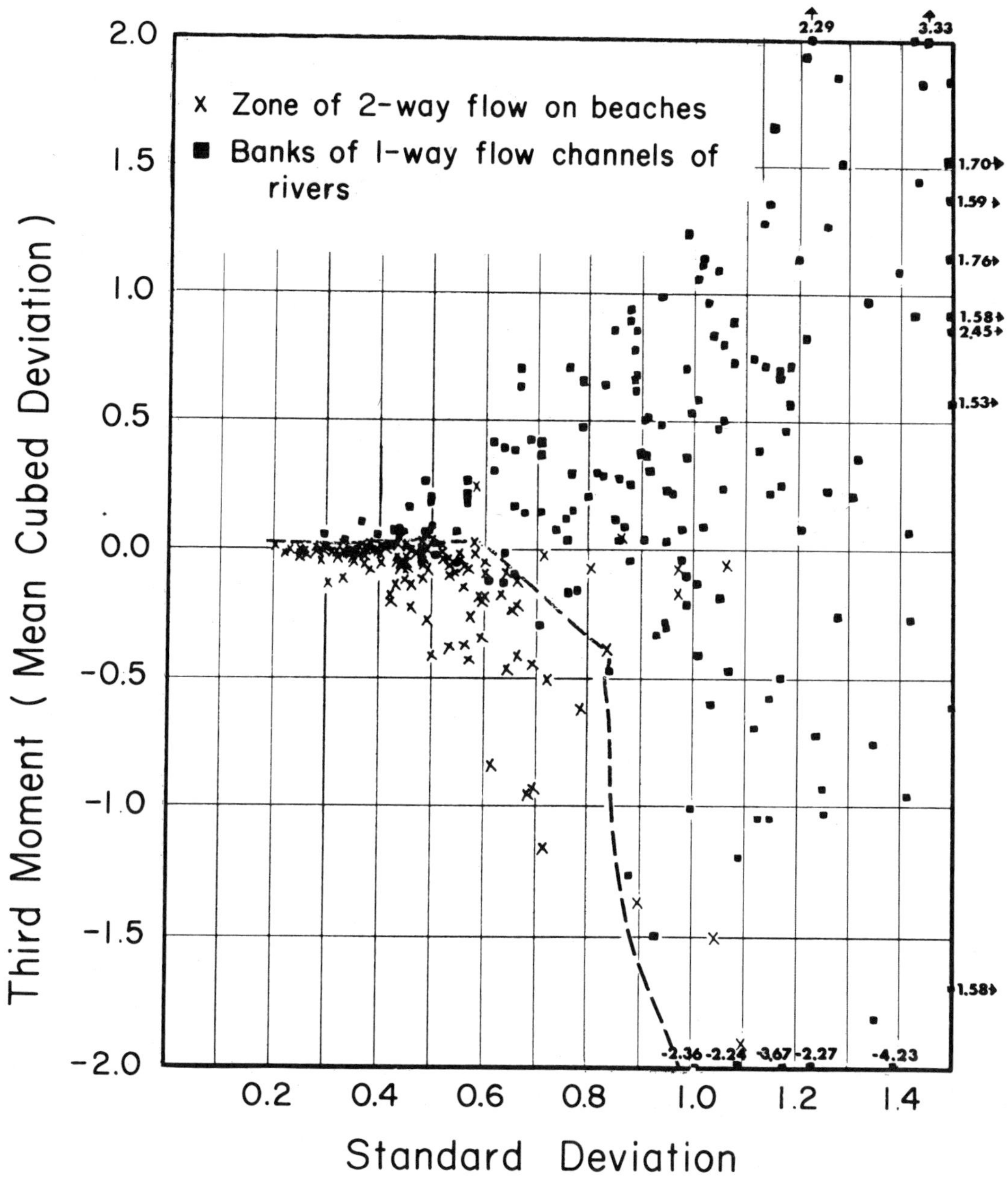

Scatter plot, standard deviation versus third moment (mean cubed deviation), for beach sands (of kind specified in text) and river sands. (G. M. Friedman, 1967, Fig. 2, p. 331, in Friedman and Sanders, 1978, Fig. 3-25, p. 81).

Figure 4-16

What stuff 'tis made of, whereof it
is born, I am to learn ....

William Shakespeare

EXERCISE 5
HEAVY-MINERAL ANALYSIS AND PROVENANCE

Introduction

In Exercise 1 we were concerned with gross mineralogical composition of sediments from the standpoint of sediment classification. Often, by studying the mineralogy in more detail we can learn much more about the history of a sediment. Specifically we can determine which rock type or types served as the parent or source material of the sediment. The source material or source area is known as provenance. It should be apparent that this is true only for detrital sediments and sedimentary rocks.

To achieve the objective of determining provenance, it is necessary to study all the minerals present in a sediment, not just the most abundant ones. Indeed it is often the least common accessory minerals that tell us most about a given sediment's provenance. Different igneous, sedimentary and metamorphic rocks have different, and often distinct or even unique, groups of accessory minerals. This laboratory will be concerned with (1) methods of separating out these accessory minerals to facilitate their identification and study and (2) consideration of questions of provenance.

Heavy Minerals

One of the more common separation methods used is based on the different specific gravities of different minerals. The more common minerals: quartz, feldspars, and calcite have low specific gravities (~ 2.7) and are known as "light minerals" or "lights." By putting a sediment sample in a liquid of high specific gravity, the common minerals will float, while the less abundant accessory minerals will sink. Since these minerals have higher specific gravities (> 2.9), they are often referred to as "heavy minerals" or "heavies."

Heavy minerals include many kinds of opaque and transparent minerals. The opaque group consists of oxides, sulfides, and other so-called ore minerals. Typically the opaque minerals predominate. The transparent group consists mostly of rock-making silicates. Table 5-1 is a list of some heavy minerals and the rocks from which they are derived.

	Parent Rocks					
	Igneous		Hydrothermal (veins, pegmatites)	Metamorphic		Sedimentary
Heavy Minerals	Mafic	Felsic		High-rank	Low-rank	
Anatase	x					
Augite	x					
Brookite	x					
Chromite	x					
Hypersthene	x					
Ilmenite	x	x				
Leucoxene	x	x				
Magnetite	x	x		x		
Olivine	x					
Rutile	x					
Serpentine	x					
Apatite		x				
Biotite		x	x	x		
Hornblende		x	x	x Blue-green variety		
Monazite		x	x			
Sphene		x				
Tourmaline		x Small pink euhedra	x Typically blue (indicolite)		x Small pale-brown euhedra, carbonaceous inclusions	
Zircon		x				
Fluorite			x			
Garnet			x	x		
Topaz			x			
Andalusite				x		
Epidote				x	x	
Kyanite				x		
Sillimanite				x		
Staurolite				x		
Zoisite				x		
Barite			x			x

SOURCE: After F. J. Pettijohn, 1975, Table 13-1, p. 487.

Some Heavy Minerals and Their Parent Rocks (Friedman and Sanders, 1978)
Table 5-1

Because they transmit polarized light, the rock-making silicates can be studied routinely as grain mounts using a binocular or petrographic microscope (Fig. 5-1). Some particles of rock-making silicate minerals have

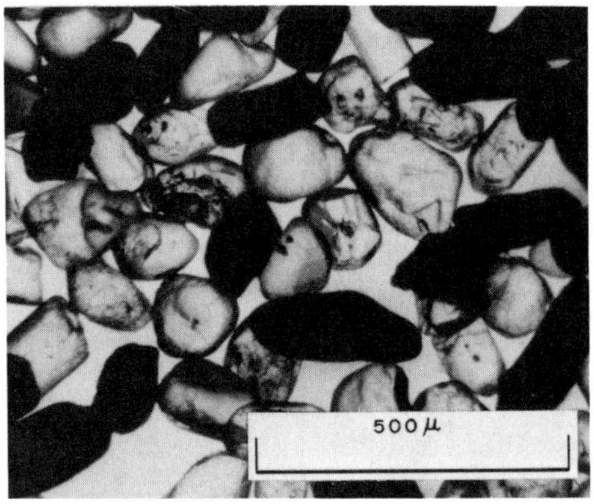

Opaque heavy minerals, including rutile and leucoxene, not distinguishable in this view (black), and transparent heavy minerals (light), all zircons. Some zircons contain many inclusions. Concentrate from Pleistocene sand attained in commercial operation by various separation techniques, viewed through binocular microscope. North Stradbrooke Island, Queensland, Australia. (Minsands Exploration Pty., Ltd., in Friedman and Sanders, 1978, Fig. 2-8, p. 37.)

Figure 5-1

been so thoroughly altered or coated that their mounted grains are no longer transparent. In rapid, routine examination of grain mounts using a petrographic microscope, the opaque minerals have been generally ignored and have been counted together under the single heading, "opaques."

Caution

The liquids used in heavy-mineral analysis are DANGEROUS! The liquids and their fumes are TOXIC, and you should carefully follow your instructor's directions. An excellent reference on the handling of these liquids is: Hauf, P. L.; and Airey, J., 1980, The handling, hazards, and maintenance of heavy liquids in the geologic laboratory: U.S. Geological Survey Circular 827, 24 p.

References Dealing with Heavy-Mineral Analysis

Andel, T. H., van, 1959, Reflections on the interpretation of heavy mineral analyses: Jour. Sedimentary Petrology, v. 29, p. 153-163.

Barsdate, R. J., 1962, Rapid heavy mineral separation: Jour. Sedimentary Petrology, v. 32, p. 608.

Beveridge, A. J., 1960, Heavy minerals in lower Tertiary formations in the Santa Cruz Mountains, California: Jour. Sedimentary Petrology, v. 30, p. 513-537.

Bramlette, M. N., 1934, Heavy mineral (sic) studies on correlation of sands at Kettleman Hills, California: Am. Assoc. Petroleum Geologists, Bull., v. 18, p. 1559-1576.

Briggs, L. I., 1965, Heavy mineral correlations and provenances: Jour. Sedimentary Petrology, v. 35, p. 939-955.

Carlston, C. W., 1946, Appalachian drainage and the highland border (sic) sediments of the Newark series: Geol. Soc. America, Bull., v. 57, p. 997-1031.

Friedman, G. M., 1961, Distinction between dune, beach and river sands from their textural characteristics: Jour. Sedimentary Petrology, v. 31, p. 514-529.

Friedman, G. M.; and Sanders, J. E., 1978, Principles of sedimentology: New York, John Wiley and Sons, 792 p., Chapter 2, Kinds of sedimentary particles (Heavy minerals, p. 35-41).

Krumbein, W. C.; and Pettijohn, F. J., 1938, Manual of sedimentary petrography: New York, NY, Appleton-Century-Crofts, Inc., 549 p., Chapter 14, Separation methods (Separation of minerals on basis of specific gravity, p. 320-344; Separation on the basis of magnetic permeability, p. 344-348).

Milner, H. B., 1962, Sedimentary petrography, 2nd ed.: v. 1, Methods in sedimentary petrography; v. 2, Principles and applications: New York, MacMillan Company, v. 1, 643 p., v. 2, 715 p.

Müller, German, 1967, Methods in sedimentary petrology: translated by Hans-Ulrich Schmincke, New York, Hafner Publishing Co., 283 p.

Sanders, J. E., 1963, Late Triassic tectonic history of the northeastern United States: Am. Jour. Sci., v. 261, p. 501-524.

Equipment and Materials

Heavy liquid - tetrabromoethane ($C_2H_2Br_4$), about 10 ml per sample

Acetone

Hand magnet, for removal of magnetic mineral particles

Separatory funnels and ring stand

Sieves

Analytical balance

Microscopes - binocular and petrographic

Index oils

Glass slides and cover glasses

Filter paper

Magnetic-separator apparatus

Bottles for collecting used heavy liquid and fluid waste

Watch glass to cover separatory funnel

Separation of Heavy Minerals

Heavy minerals may be separated from minerals of lower density by (1) gravity methods, (2) magnetic methods, or (3) centrifuge methods. In this exercise you will carry out procedures that will introduce you to the principles of gravity separation and magnetic separation.

Separation of heavy-mineral particles by gravity is accomplished using a so-called heavy liquid and two superposed separatory funnels. The technique is reliable and particularly useful where large amounts of sediment must be processed. It suffers, however, from the disadvantage that complete separations normally require from one to three hours. A typical arrangement of apparatus for gravity separation of heavy minerals is shown in Figure 5-2.

In order to serve as a suitable medium for gravity separation of heavy-mineral particles, a heavy liquid, in addition to having a high specific gravity, must be: liquid at room temperature, of low-viscosity, transparent, chemically inert, readily available, chemically stable, easily handled and recovered, inexpensive and non-poisonous. Unfortunately, heavy liquids having the last two properties are not common.

You should keep in mind at all times that THE LIQUID THAT YOU WILL BE USING IS HIGHLY TOXIC AND THAT THE FUMES ARE TOXIC AS WELL. ALL

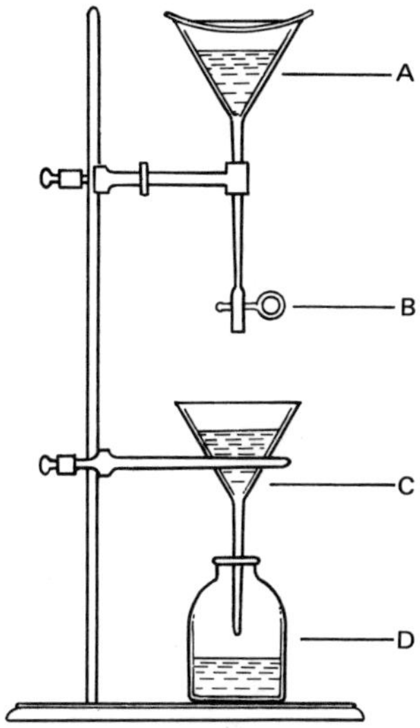

Simple Apparatus for Heavy-Mineral Separation by Gravity Method

A. Separatory funnel containing heavy liquid

B. Rubber tube with pinch-cock

C. Funnel with filter-paper lining

D. Bottle for collecting used heavy liquid

(Krumbein and Pettijohn, 1938, Fig. 153, p. 335.)

Figure 5-2

PROCEDURES MUST BE CARRIED OUT UNDER A WELL-VENTILATED HOOD! You will use tetrabromoethane ($C_2H_2Br_4$), which has a specific gravity at 20°C of 2.96; bromoform ($CHBr_3$) is also commonly used as a heavy liquid. Although for this exercise you will perform only one separation, you should be aware that using different liquids or mixtures of liquids of proper densities one can separate heavies into smaller and smaller classes of differing specific gravity.

Procedure I: Separation of Heavy-Mineral Particles by the Gravity Method

1. If necessary, sieve sample in order to obtain material in 1 to 4ϕ (1/2-1/16 mm) range.

2. Weigh sample - about 25 g is sufficient. Record sample weight, heavy liquid and liquid specific gravity on Form 5-1.

3. Pour tetrabromoethane into upper (separatory) funnel (see Fig. 5-2). The funnel should be about 1/2 full. CARRY OUT THIS AND ALL FOLLOWING OPERATIONS UNDER A WELL-VENTILATED

HOOD. THE HEAVY LIQUID AND ITS FUMES ARE TOXIC!

4. Pour sample into tetrabromoethane and stir thoroughly in order to wet all particles and disperse air bubbles.

5. Allow particles to settle; stir periodically so that particles will not adhere to funnel wall. Cover funnel with watch glass to reduce heavy-liquid evaporation loss.

6. When heavy minerals have settled to bottom of separatory funnel, open pinch-cock and allow heavy-mineral particles to drop onto filter paper in lower funnel. Close pinch-cock so that minerals floating in remaining heavy liquid will remain in the separatory funnel.

7. After heavy liquid has drained from filter paper into used heavy-liquid bottle below, remove paper and place upside down in porcelain dish or large watch glass containing acetone. If necessary, use plastic squeeze bottle containing acetone to wash into the dish any particles which adhere to the filter paper; decant excess acetone into fluid waste container.

8. Dry heavy-mineral fraction, record weight on Form 5-1, and label and retain fraction for compositional determination and provenance analysis.

9. Using the same apparatus set-up as before, drain the remaining heavy liquid through filter paper in the lower funnel and into the used heavy-liquid bottle below. The light minerals will collect on the filter paper. It may be necessary to use a squeeze bottle containing acetone to wash particles that adhere to the side of the separatory funnel. If this is necessary, it is important during the washing process to replace the bottle used for collection of undiluted used heavy liquid with one used for collection of fluid waste (i.e., a mixture of acetone and heavy liquid).

10. Using the same procedure as that outlined in Steps 7 through 9, collect, dry, record weight and retain the light-mineral fraction for compositional determination.

HEAVY-MINERAL SEPARATION BY GRAVITY METHOD

Sample weight ______________________________

Type of heavy liquid ________________________

Heavy liquid specific gravity _________________

Weight of heavy-mineral fraction _______________

Weight of light-mineral fraction _______________

Form 5-1

Magnetic separation of mineral particles depends on the magnetic susceptibility of the minerals to be separated. This is a complex phenomenon which is a function of chemical composition, especially minor amounts of iron or manganese, and atomic-lattice structure.

The simplest form of magnetic separation is by hand magnet, a procedure which works for only one common detrital mineral - magnetite. Other minerals having magnetic susceptibility are separated by using apparatus which generate strong electromagnetic fields. Companies which market electromagnetic separators suitable for mineralogic work are: S. G. Frantz Co., Trenton, New Jersey; Chas. W. Cook & Sons, Ltd., Birmingham, England; and R. Eisenblaetter, Lingolsheim (Bas-Rhin), France.

The Frantz separator (Fig. 5-3) incorporates an electromagnet with two elongate pole pieces arranged so that the space between the poles is much wider on one side than the other. A vibrating metal chute parallels the pole pieces. Mineral particles are introduced into the upper end of the chute and slide toward the lower end. Those with higher magnetic susceptibility move toward the side of the chute where the pole gap is

Frantz Isodynamic Magnetic Separator
(inclined position)

(Müller, 1967)

Figure 5-3

narrow and the magnetic flux the greatest. The separator is mounted so that it can be rotated both in the direction of grain movement (slope) and in a direction normal to the direction of grain movement (tilt). The way in which mineral particles separate as they move along the length of the chute depends on (1) the tilt of the chute, (2) the amperage applied to the electromagnet and, to some extent, (3) the slope and rate of feed to the chute. At the lower end of the chute the particles are separated into two streams, one consisting of grains of higher susceptibility than that corresponding to the amperage setting used, the other consisting of grains of lower susceptibility.

Inasmuch as there is wide variation in magnetic susceptibility of mineral particles of a given species, it is common practice to set the Frantz separator slope at 10 to 30° (20° is more or less standard), the tilt at 5 to 20° (5° for minerals of low susceptibility, 15° for those of moderate to high susceptibility) and then determine the most effective amperage by trial and error.

The magnetic separator manufactured by Chas. W. Cook & Sons is very similar to the Frantz separator; the Eisenblaetter separator, developed in the early 1960s, has the advantage that in a single operation sample material is separated during free fall into fractions having different mass susceptibilities. The procedure that follows is designed for use with the Frantz apparatus, but could be easily modified for use with the Cook separator.

Procedure II: Separation of Heavy-Mineral Particles by the Magnetic-Susceptibility Method

1. Before using the electromagnetic separator, remove ferromagnetic minerals (magnetite and pyrrhotite) from the sample with a hand magnet. These minerals will clog the apparatus if left in the sample.

2. Using a side slope of 20° and a tilt of 25°, run the sample at a setting of 0.4 amperes. Label and retain the magnetic fraction.

3. Take the nonmagnetic fraction from Step 2 and run it at a setting of 0.8 amperes. Label and retain the magnetic fraction.

4. Take the nonmagnetic fraction from Step 3 and run it at a setting of 1.5 amperes. Label and retain the magnetic fraction and the nonmagnetic fraction.

5. Refer to Table 5-2 for a listing of those heavy minerals which are magnetic at 0.4, 0.8, and 1.5 amperes.

Magnetic at 0.4	Magnetic at 0.8	Magnetic at 1.5	Nonmagnetic at 1.5
Garnet	Biotite	Muscovite	Zircon
Ilmenite	Hornblende	Spinel	Rutile
Chromite	Hypersthene	Enstatite	Sphene
Chloritoid	Augite	Tourmaline	Leucoxene
Olivine	Actinolite	Clinozoisite	Apatite
	Staurolite	Diopside	Corundum
	Epidote	Tremolite	Barite
	Chlorite		Fluorite
			Sillimanite
			Kyanite

Minerals which are magentic at 0.4, 0.8, and 1.5 amperes settings (side slope 20°, tilt 25°) on the Frantz Isodynamic Magnetic Separator

Table 5-2

Determination of Provenance

Based on their physical and chemical nature, heavy minerals may be divided into four groups: opaques, micas, ultra-stables and meta-stables (Folk, 1974, p. 96).

Opaque minerals, inasmuch as they have a relatively high iron content, generally have a very high specific gravity. They include magnetite, ilmenite, pyrite, hematite, limonite, and leucoxene. Most researchers lump the opaques together and do not attempt to differentiate them. However, for purposes of this exercise you should attempt differentiation.

Micas vary in specific gravity due to compositional variation and in hydraulic behavior due to variation in particle shape. In addition, mica particles sometimes do not sink in bromoform. It is understandable that the micas are commonly not counted in heavy-mineral analyses.

The ultra-stable group includes zircon, tourmaline and rutile. Zircon and tourmaline are chemically inert and very hard and are therefore able to survive numerous cycles of erosion and transport. They are also polyvarietal and thus excellent indicators of provenance. Euhedral zircon is an indicator of volcanism.

The metastable group includes olivine, apatite, amphibole, pyroxene, garnet, epidote, clinozoisite, zoisite, kyanite, sillimanite, andalusite, staurolite, and many other minerals.

Table 5-3 is a summary of the characteristics and mode of occurrence of the more important heavy minerals. It will be useful in Procedure III: Identification of Heavy Mineral Particles. For detailed information on identification of both heavy minerals and light minerals, refer to Krumbein and Pettijohn (1938, p. 412-464), or Milner (1962).

Procedure III: Determination of Heavy-Mineral Composition

1. Under the binocular microscope, study the composition of the heavy-mineral fraction that you saved from Procedure I: Separation of Heavy-Mineral Particles by the Gravity Method.

2. Estimate particle-type percentages and record data on Form 5-2.

3. Using the fractions that you saved from Procedure II: Separation of Heavy-Mineral Particles by the Magnetic Susceptibility Method, estimate particle-type percentages and record data on Form 5-2.

4. Diagram your estimates of the percentages of various particle-types on Form 5-2.

The data that you have recorded in Procedure III will be used in Procedure IV: Interpretation of Provenance.

In order to interpret the nature of the source rock from which sediment was derived, it is necessary to group heavy minerals into genetic suites such as (1) reworked sedimentary, (2) low- and high-rank metamorphic, (3) sialic and mafic igneous, (4) pegmatitic, and (5) authigenic (Hubert, 1971, p. 462). The relative importance of each suite is then evaluated and inferences made about provenance. Table 5-4 outlines suites of heavy minerals derived from various source rocks.

Table 5-3

Characteristics and Mode of Occurrence of Important Heavy Minerals
(After Krumbein and Pettijohn, 1938)

Mineral	G	H	Form	Colors	Luster	Miscellaneous	Provenance
<u>Opaques</u>:							
Magnetite (Fe_3O_4)	5.17	6	Isometric, octahedra	Bluish black in reflected light	Metallic	Angular and well rounded particles abundant; marked by strong magnetic character and crystal form; distinguished from ilmenite with difficulty; may alter to hematite or limonite	Mafic igneous and high-rank metamorphic rocks
Ilmenite ($FeTiO_3$)	4.6-4.9	5-6	Orthorhombic	Brownish to purplish black in reflected light	Metallic	Common in sediments as irregular to well-rounded grains; often in part altered to leucoxene; nonmagnetic	Igneous and high-rank metamorphic rocks
Pyrite (FeS_2)	5.02	6	Isometric, cubes and pyritohedra, often striated; conchoidal fracture	Pale brass-yellow in reflected light, readily tarnished	Metallic	Nearly always authigenic; occurs in sediments as globular to irregular aggregates or as nodules and small concretions	Ubiquitous

Table 5-3 (continued)

Mineral	G	H	Form	Colors	Luster	Miscellaneous	Provenance
Hematite (Fe_2O_3)	5.2	6	Hexagonal-rhombohedral	Indian red to black in reflected light	Metallic to earthy	Usually alteration product but may be detrital	Ubiquitous
Limonite Hydrous iron oxide	3.8	5-6	Amorphous	Ochre yellow, brown to brownish black in reflected light	Earthy to metallic	Occurs as rounded granules or as powdery aggregates and coatings, usually as a decomposition product of pyrite, marcasite and glauconite; sometimes detrital	Ubiquitous
Leucoxene	3.5-4.5		An aggregate of extremely fine-grained sphene, rutile or anatase	Dead white in reflected light	Dull	Forms an alteration product usually after ilmenite; appears as rounded grains, sometimes with mat surface minutely pitted	Authigenic
<u>Micas</u>:						Do not always sink in bromoform; percentages unreliable because of unpredictable hydraulic behavior	

Table 5-3 (continued)

Mineral	G	H	Form	Colors	Luster	Miscellaneous	Provenance
Ultra-stables:							
Zircon ($ZrSiO_4$)	4.6-4.7	7½	Tetragonal, short prisms with pyramids; cleavage rare	Usually colorless, some particles are mauve, yellow to brown	Adamantine	Particles often show euhedral form with well-marked crystal faces; well-worn particles usually elongate, elliptical or globular; contain large mineral, liquid or gas inclusions; many particles show well-defined zoning; particles can survive many reworkings	Sialic to intermediate igneous rocks
Tourmaline	3.0-3.3	7-7½	Hexagonal-rhombohedral; striated prismatic; cleavage lacking or poor	Yellow-brown, dark brown, indigo to black		Particles occur as elongate prisms, irregular fractured pieces and well-rounded ovals; inclusions are common. Particles can survive many reworkings	Pneumatolytic rocks, pegmatites, schists, gneisses, marbles
Rutile (TiO_2)	4.24	6-6½	Tetragonal, often striated	Yellow, reddish brown, red	Adamantine	Particles irregular, generally elongate, prismatic with rounded pyramidal ends common; inclusions abundant. Characterized by form, very high bire-	Sialic igneous and crystalline metamorphic rocks

Table 5-3 (continued)

Mineral	G	H	Form	Colors	Luster	Miscellaneous	Provenance
Rutile (continued)						fringence, high relief, deep color and striae	
<u>Metastables</u>:							
Olivine $(Mg, Fe)_2SiO_4$	3.5	6½-7	Orthorhombic, cleavage good	Colorless, pale yellow	Glassy	Occurs in sediments as irregular, much-fractured particles; particles colorless, with bright interference colors and showing traces of decomposition; rare as a detrital occurring only in dry climates with rapid erosion	Basic and ultrabasic igneous rocks
Apatite $Ca_5(F, Cl)(PO_4)_3$	3.16-3.22	5	Hexagonal, long to short prismatic, terminated by base or pyramid; imperfect basal cleavage	Green or brown, also blue violet or colorless	Vitreous to subresinous	Particles oval to nearly circular to slightly worn elongate prismatic form; frequently contain inclusions in planes or rows; characterized by detrital form, low birefringence	Acid igneous rocks and pegmatites

Table 5-3 (continued)

Mineral	G	H	Form	Colors	Luster	Miscellaneous	Provenance
Amphibole (Hornblende, Actinolite, Tremolite)	2.9-3.5	5-6	Long, blade-like prisms; cleavage-two oblique	Hornblende-dark brown and green to black; Actinolite-dusky yellow green to dusky yellowish green; Tremolite-white to colorless	Satiny, glassy, pearly	Particles elongate, prismatic, with longitudinal cleavage and marked diagonal cross-fractures; may occur as irregular fractured particles; hornblendes often nearly opaque and appear translucent just on thin edges	Igneous and metamorphic rocks
Pyroxene (Augite, diopside)	3.2-3.4	5-6	Monoclinic; cleavage-perfect prismatic at about 90°	Augite-pale brownish gray or pale grayish green; Diopside-very pale green or colorless	Vitreous	Particles usually elongate, worn cleavage fragments, sometimes with dentate ends; irregular or poorly rounded; diopside more resistant than augite, particles moderately well-rounded	Augite-intermediate and basic igneous rocks; Diopside-metamorphic contact rocks and schists
Garnet	3.5-4.3	7	Isometric, dodecahedra, trapezohedra; fracture-conchoidal	Colorless, pale pink, orange, red, apricot-yellow, amber	Glassy	Characterized by color, conchoidal fracture, isotropism and high relief	Igneous and metamorphic rocks; high abundance indicates metamorphic source

Table 5-3 (continued)

Mineral	G	H	Form	Colors	Luster	Miscellaneous	Provenance
Epidote $Ca_2(Al,Fe)$ $Al_2(Si_2O_7)$ $(SiO_4)O(OH)$	3.36	6	Monoclinic; cleavage-present but not visible in particles	Pale greenish yellow to lemon yellow	Glassy to dull	Particles greenish yellow, equidimensional, sharply angular to subrounded	Metamorphosed igneous rocks
Clinozoisite (Complex calcium aluminum silicate)	3.25-3.37	6-6½	Monoclinic; prismatic	Grayish white, green and pink; Lighter in color than epidote	Vitreous	Zoisite is a relatively rare member of the epidote group	Crystalline schists and metamorphosed basic rocks
Kyanite (Al_2SiO_5)	3.6	5-7	Triclinic; bladed; cleavage-2 at 90°	Usually colorless, rarely pale blue	Vitreous to pearly	Particles are elongate and of marked rectangular outline to short, moderately rounded and elliptical; carbonaceous inclusions common; particles may alter along edges to "mica"	Schists and gneisses
Sillimanite Al_2SiO_4	3.23	6-7	Orthorhombic; cleavage - one direction, perfect	Colorless	Vitreous	Particles irregular to short prismatic, split longitudinally and have striae parallel to length	Metamorphosed argillaceous rocks

Table 5-3 (continued)

Mineral	G	H	Form	Colors	Luster	Miscellaneous	Provenance
Staurolite $(Fe,Mg)_2(Al,Fe)_9O_6(SiO_4)_4(O,OH)_2$	3.7	7-7½	Orthorhombic; short prisms, cruciform twins	Yellow, gold, brown	When altered, dull to earthy; when fresh, resinous to vitreous	Particles irregular, somewhat platy with hackly to subconchoidal fracture; inclusions numerous; bright interference colors	Crystalline schists, slates and sometimes gneisses
Andalusite (Al_2SiO_5)	3.15	7½	Orthorhombic; nearly square prisms; cleavage-good prismatic at about 90°	Colorless to pink	Dull to vitreous	Particles are elongate worn or broken prisms to irregular in shape showing only conchoidal fracture; inclusions common	Contact metamorphic zones in shaly rock

DETERMINATION OF HEAVY-MINERAL COMPOSITION - Gravity separated fraction and magnetically separated fraction

Exercise 5
Name ______________________
Laboratory Section ______________
Date ______________________

	Particle-type	Gravity-separated	Magnetically-separated
Estimated Percentage of Total Particles	Magnetite		
	Ilmenite		
	Pyrite		
	Hematite		
	Limonite		
	Leucoxene		
	Total Opaques		
	Total Micas		
	Zircon		
	Tourmaline		
	Rutile		
	Total Ultra-Stables		
	Olivine		
	Apatite		
	Amphiboles		
	Pyroxenes		
	Garnet		
	Epidote		
	Clinozoisite		
	Kyanite		
	Sillimanite		
	Staurolite		
	Andalusite		
	Total Metastables		
	Total Particles	100%	100%
	Pie Diagrams (3.6° of arc = 1% composition)	Heavy mineral composition	Heavy mineral composition

Form 5-2

Reworked sediments	Well-rounded grains of rutile, tourmaline, zircon.
Low-rank metamorphic	Biotite, chlorite, spessartite garnet, tourmaline (especially small, euhedral, brown crystals with graphite inclusions).
High-rank metamorphic	Actinolite, andalusite, apatite, almandine garnet, biotite, diopside, epidote, clinozoisite, glaucophane, hornblende (including blue-green varieties), ilmenite, kyanite, magnetite, sillimanite, sphene, staurolite, tourmaline, tremolite, zircon.
Sialic igneous	Apatite, biotite, hornblende, ilmenite, monazite, muscovite, rutile, sphene, tourmaline, zircon.
Mafic igneous	Augite, diopside, epidote, hornblende, hypersthene, ilmenite, magnetite, olivine, oxyhornblende, pyrope garnet, serpentine.
Pegmatites	Apatite, biotite, cassiterite, garnet, monazite, muscovite, rutile, tourmaline (especially indicolite).
Ash falls	Euhedral crystals of apatite, augite, biotite, hornblende, and zircon.
Authigenic	Hematite, leucoxene, limonite, tourmaline, zircon; euhedral crystals of anatase, brookite, pyrite, rutile, and sphene.

Provenance of Some Common Heavy Minerals
(Hubert, 1971, p. 462)

Table 5-4

Procedure IV: Interpretation of Provenance

The data on Form 5-2 provide a basis for interpretation of provenance for (1) the gravity-separated heavy-mineral fractions, and (2) the magnetically-separated heavy-mineral fraction. In the space below, outline your interpretations and the reasoning that led you to them:

EXERCISE 6

Science is a great many things, but
in the end they all return to this:
science is the acceptance of what
works and the rejection of what
does not ...

Bronowski

EXERCISE 6
CARBONATE STAINING AND PEELS

Introduction

In Exercise 1, SEDIMENTARY PARTICLES, you stained sediments in order to differentiate quartz and feldspar and to identify calcite. The purpose of Exercise 6 is to provide further introduction to carbonate staining procedures and also an introduction to techniques of peel preparation.

Staining is one of the most useful techniques in the analysis of sedimentary rocks, particularly in carbonate analysis. It is indispensable in identifying the minerals that make up limestones and dolostones, and hence serves as a useful tool in fabric analysis. It can be used in modern carbonate sediments and in partially or wholly consolidated Pleistocene rocks to identify the metastable carbonate minerals aragonite and magnesium calcite. It is equally useful in the differentiation of calcite, dolomite, gypsum, and anhydrite in modern and ancient limestones, dolostones, and evaporites. Identification of carbonate minerals is effective in hand specimen, polished surface, and thin-section; in addition, stained peels of carbonate rocks can be prepared simply and effectively.

Staining in the study of carbonate rocks dates back to the classical work of Lemberg (1887), an extensive literature dealing with carbonate staining having developed since then. The following list of references includes some of the more important ones from the standpoint of both theory and practical applications in staining and peel preparation.

References Dealing with Carbonate Staining and Peels

Choquette, P. W.; and Trusell, F. C., 1978, A procedure for making the titan-yellow stain for Mg-calcite permanent: Jour. Sedimentary Petrology, v. 48, n. 2, p. 639-641.

Davies, P. J.; and Till, R., 1968, Stained dry cellulose peels of ancient and Recent impregnated carbonate sediments: Jour. Sedimentary Petrology, v. 38, p. 234-237.

Evamy, B. D., 1963, The application of a chemical staining technique to a study of dolomitization: Sedimentology, v. 2, p. 164-170.

Feigl, Fritz, 1937, Qualitative analysis by spot tests: New York, Nordemann Publishing Company, 400 p.

__________, 1958, Spot tests in inorganic analysis, 5th ed.: New York, Elsevier Publishing Co., 600 p.

Frank, Ruben M., 1965, An improved carbonate peel technique for high powered studies: Jour. Sedimentary Petrology, v. 35, p. 499-500.

Friedman, G. M., 1959, Identification of carbonate minerals by staining methods: Jour. Sedimentary Petrology, v. 29, p. 87-97.

__________, 1971, Staining, Ch. 22, p. 511-530, in Carver, R. E., ed., Procedures in sedimentary petrology: New York, Wiley-Interscience, 653 p. (Staining of carbonates, p. 513-522).

__________, 1977, Identification by staining methods of minerals in carbonate rocks, p. 96-97, in LeRoy, L. W.; LeRoy, D. O.; and Raese, J. W., Subsurface geology, 4th ed., Colorado School of Mines, Golden, 941 p.

Ives, William, Jr., 1955, Evaluation of acid etching of limestone: Kansas Geol. Surv. Bull. 114, pt. 1.

Katz, Amitai; and Friedman, G. M., 1965, The preparation of stained acetate peels for the study of carbonate rocks: Jour. Sedimentary Petrology, v. 35, p. 248-249.

Lamar, J. E., 1950, Acid-etching in the study of limestones and dolomites: Illinois State Geol. Surv. Circ. 156, 47 p.

Lemberg, J., 1887, Zur microchemischen Untersuchung von Calcit, Dolomit und Predazzit: Zeits. Deutschen Geologischen Ges., v. 39, p. 489-492.

Warne, Slade, 1962, A quick field or laboratory staining scheme for the differentiation of the major carbonate minerals: Jour. Sedimentary Petrology, v. 32, p. 29-38.

Wolf, K. H.; and Warne, Slade, 1960, Remarks on the application of Friedman's staining methods: Jour. Sedimentary Petrology, v. 30, p. 496-497.

Equipment and Materials

Cutting and polishing materials and equipment for rock-slab preparation

Laboratory sink and running water

Beakers and Bunsen burner

Binocular microscope

Hand lens
Dilute hydrochloric acid
Sodium hydroxide
Acetate film; clear, 0.005 gauge
Acetone
Stains:
- Alizarine red S
- Feigl's solution
- Titan yellow or clayton yellow

Procedures

Many stains have been developed to promote identification of carbonate minerals; some are more effective and some "take" more rapidly than others. Staining schemes usually depend on the differences in rate of solution of carbonate minerals, for instance, between aragonite and calcite and between calcite and dolomite. Continuing exposure to a stain ultimately affects both minerals that one wishes to distinguish; hence the reaction has to be stopped before both are stained. Effective reaction time depends on the stain and on the mineralogy and fabric of the rock, and varies from one stain to another and often from one rock to another. Although staining techniques are easy, the beginner has to learn each technique and develop a feel for it. If one expects excellent results on the first try, disappointment often results.

In general, the procedures that you will follow in this exercise combine the techniques of (1) carbonate-mineral differentiation by staining, and (2) analysis of rock texture and fabric with the aid of acetate film peels. Specifically, the procedures are I: Slab Preparation; II: Etching the Slab; III: Staining the Slab; and IV: Preparation of the Peel.

Procedure I: Slab Preparation

1. Saw carbonate rock specimen(s) on trim saw to obtain surface(s) approximately 8 x 8 cm.
2. Grind surface(s) with carborundum powder; final-polish with 800 grade powder.
3. Thoroughly wash polished surface to provide clean surface for etching procedure.

Procedure II: Etching the Slab
(Etching is the most important single step in preparing peels of good quality.)

1. Prepare acid solution by using eight to ten parts HCL (commercial grade) diluted with water to 100 parts. (Lamar, 1950, p. 2; Ives, 1955, p. 8.)

2. Expose polished surface(s) to acid solution 2 to 3 minutes, depending on sample fabric and mineralogy. Wash surface(s) with running water to stop etching process.

3. Examine surface(s) under hand lens or binocular microscope to make sure that relief has developed. A well-etched sample appears in three dimensions under a binocular microscope.

The staining scheme that you will use for this exercise is in general use for routine differentiation of carbonate minerals (Friedman, 1959). It is outlined in Figure 6-1. It permits differentiation of dolomite, calcite, aragonite, magnesium-calcite, gypsum, and anhydrite.

Although Warne (1962) expanded Friedman's staining scheme for carbonate minerals to witherite, rhodochrosite, smithsonite, cerrusite, and siderite, staining procedures in this exercise are limited to those for the common minerals of carbonate sediments, limestones, dolostones, and evaporites. Carbonate and sulfate minerals subject to stain analysis are listed in Table 6-1.

Mineral	Composition
Calcite	$CaCO_3$ (Hexagonal)
Aragonite	$CaCO_3$ (Orthorhombic)
Dolomite	$CaMg(CO_3)_2$
Magnesite	$MgCO_3$
High-Magnesian Calcite	Solid Solution of $MgCO_3$ in calcite
Gypsum	$CaSO_4 \cdot 2H_2O$
Anhydrite	$CaSO_4$

Carbonate and Sulfate Minerals Subject to Stain Analysis

Table 6-1

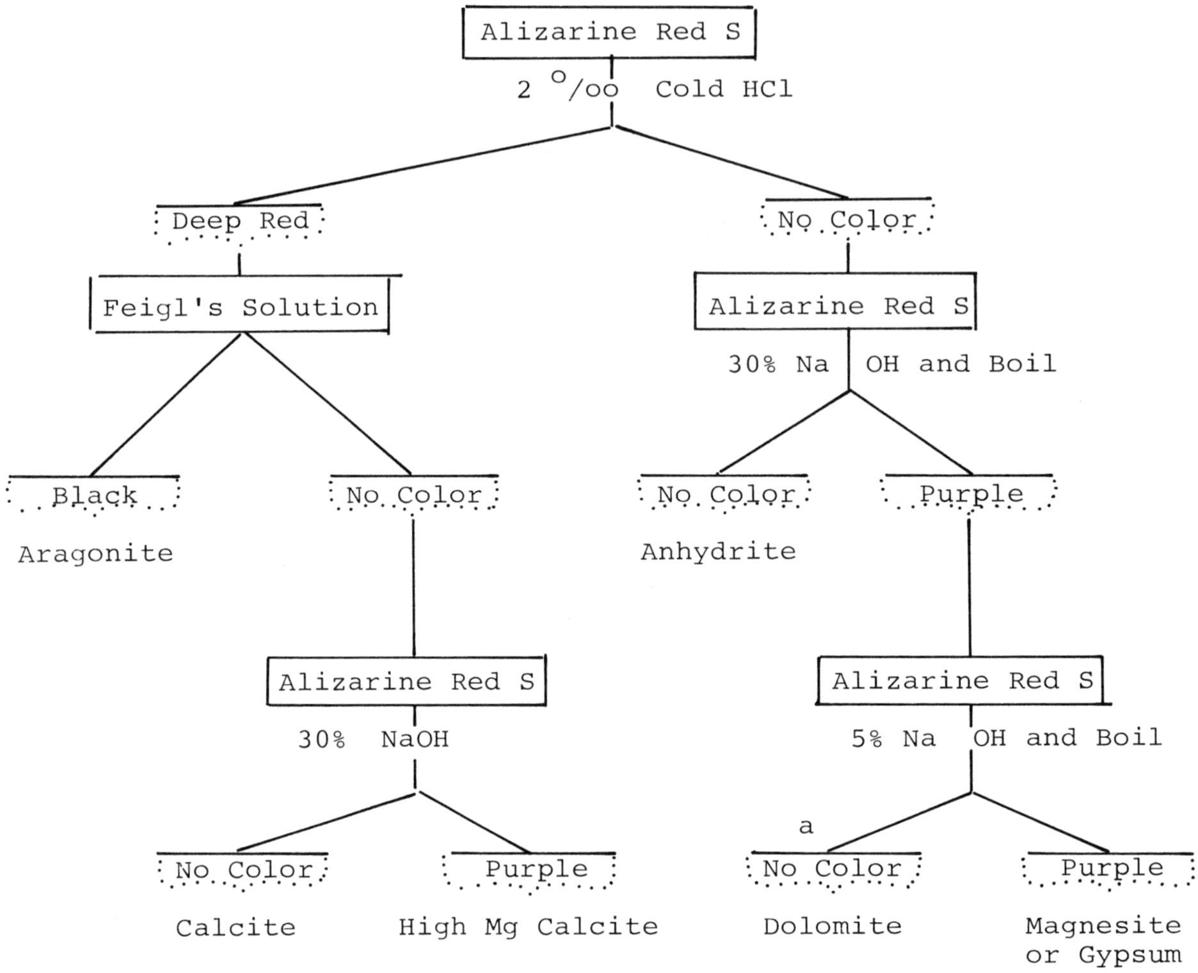

Recommended staining procedure; alizarine red S and Feigl's solution. [a]Or faint stain. (Friedman, 1971).

Figure 6-1

The minerals listed in Table 6-1 react differently to various organic and inorganic stains, the most important of which are described in the following few paragraphs. For more complete information about stain characteristics and preparation, refer to Choquette (1978), Evamy (1963), Friedman (1959, 1971), Warne (1962), and Wolf and Warne (1960).

Alizarine red S, an organic stain, is the most effective stain for calcite. It is prepared by dissolving 0.1 gm in 100 ml 0.2%

hydrochloric acid (add 2 ml of commercial grade concentrated hydrochloric acid to 998 ml of water). Calcite is stained deep red within 2 to 3 minutes, whereas dolomite is not stained, except on excessive exposure. However, staining time should be considered flexible, depending on the texture of the rock studied. For some studies a half-minute exposure is ample; it results in a pale red coloration of the calcite.

Alizarine red S is suited particularly for thin-sections. Staining time varies slightly, but is usually no longer than a minute. Staining of the thin-section should continue until sufficient contrast between calcite, which is stained red, and other minerals, such as dolomite, which remain unstained, is developed.

Titan yellow staining solution is prepared by dissolving 0.2 g of Titan yellow in 25 cc of methanol. Add 15 cc of 30% NaOH and boil. The solution stains dolomite a yellowish color, while calcite is unaffected. Other stains, especially Alizarine cyanine green, may be substituted for Titan yellow (see Friedman, 1959, Table 3). Clayton yellow may also be substituted for Titan yellow (see Choquette, 1978).

Of several stains specific for aragonite the most sensitive was developed by Feigl (1937, p. 328-329). Friedman (1959) named this reagent "Feigl's solution." The preparation of this solution is as follows:

> 1 gm of solid (commercial grade) Ag_2SO_4 is added to a solution of 11.8 gm $MnSO_4 \cdot 7H_2O$ in 100 ml of water and boiled. After cooling the suspension is filtered and one or two drops of diluted sodium hydroxide solution is added. The precipitate is filtered off after 1 or 2 hours (Feigl, 1958, p. 470). It is important that only distilled water be used; tap water leaves a white precipitate of silver chloride (Katz and Friedman, 1965).

Feigl's solution turns aragonite black but does not effect calcite.

It is worth noting that (1) the dyes which stain calcite in acid solution stain dolomite (and magnesite) in alkaline solution and that (2) Alizarine red S gives the most satisfactory general purpose results of the organic dyes; they do not require boiling, and their action is fast and reliable.

In the following procedure you will differentiate calcite, dolomite and aragonite using the staining scheme recommended by Friedman (1971). Your instructor will provide the prepared staining solutions that you will need.

Procedure III: Staining the Slab

1. The slab(s) etched and washed under running water in Procedure II should be dry by now.

2. Drop Alizarine red S on the polished surface(s) and allow it to stand for 4 minutes.

3. Gently rinse with water and examine surface(s) under the binocular microscope. Those mineral particles which stain deep red are either calcite or aragonite; those which do not accept stain are anhydrite or dolomite.

4. Continue testing for mineralogical composition using the scheme outlined in Figure 6-1. Your instructor will provide specific instructions regarding (a) the number of slabs you should test, and (b) the extent to which you should complete the entire sequence of tests.

Peels are very useful in providing information on the texture and fabric of carbonate rocks. In addition, if the slab is stained, mineralogical composition can be determined. Since for stained peels the actual layer of stain is transferred from the polished rock surface to the peel, it is important to have the best possible "take" of the stain on the slab. For this reason, staining is accomplished by total immersion of the slab in the solution.

Peels are made using (1) commercially available acetate sheets, (2) plexiglass, or (3) special peel sheets prepared in the laboratory. In the following procedure, you will use acetate sheets.

Procedure IV: Preparation of the Peel

1. Staining:

    a. For all stains except dolomite stain, immerse the etched rock slab in staining solution, polished side up. Staining takes about 4 minutes, but some experimentation may be required. Pour off solution and wash slab with water (for Feigl's solution only distilled water should be used).

    b. When staining for dolomite, immerse etched rock slab in boiling solution for 5 to 10 minutes.

2. Immerse slab in acetone solvent for not more than 2 or 3 seconds.

3. Immediately place a piece of acetate sheet on etched, stained surface. Press firmly and evenly with fingers so that no air bubbles remain under the sheet.

4. Leave slab with attached acetate sheet to dry for about 40 minutes.

5. Carefully remove acetate sheet from slab.

6. Stained peels may be mounted between glass slides for preservation and permanent storage. If 35 mm glass slide mountings are used, in addition to the storing advantage, the peels may be projected onto a large screen for convenient group examination and study.

Study the rock peels and on Form 6-1 record observational data on (1) texture, (2) fabric, and (3) mineralogical composition. Retain the peels for reference when you do Exercise 7 - Description and Classification of Sedimentary Rocks in Hand Specimen and Thin Section: Intrabasinal Rocks.

STAINED CARBONATE-ROCK PEELS - Observational data on texture, fabric and mineralogical composition

Exercise 6
Name ______________________
Laboratory Section ______________
Date ______________________

Peel No.	Observational Data		
	Texture	Fabric	Composition

Form 6-1

EXERCISE 7

...... the nature and system of
the world have been discovered
but lately ......

Lucretius

EXERCISE 7
DESCRIPTION AND CLASSIFICATION OF SEDIMENTARY ROCKS IN HAND SPECIMEN AND THIN SECTION: INTRABASINAL ROCKS

Introduction

Sedimentary rocks may be divided into five major categories: (1) carbonate rocks, (2) authigenic rocks, (3) carbonaceous rocks, (4) terrigenous rocks, and (5) pyroclastic rocks. These five categories of rocks differ in mineral and chemical composition, as well as in origin. Carbonate rocks consist mostly of calcite or dolomite; authigenic rocks are of various kinds and include evaporites, such as sulfate and halide minerals, and other chemical rocks, such as cherts, and iron-rich and manganese-rich rocks; carbonaceous rocks consist of carbonaceous matter; terrigenous rocks of various silicate minerals derived from land; and pyroclastic rocks of various silicate minerals and glasses derived from volcanoes. Rocks of the first three categories originate and for the most part are preserved in the basin of deposition. These are grouped together and considered intrabasinal rocks, that is, sedimentary rocks whose constituent particles were derived from within the basin of deposition. Thus, most carbonate sediments are secreted or were precipitated and are preserved within a basin of deposition. Although tree logs can travel some distance, most carbonaceous rocks, such as coal, are deposited and preserved _in situ_. By contrast the terrigenous rocks belong to the group designated extrabasinal rocks, which includes sedimentary rocks composed of particles derived from outside the basin of deposition. Such particles may travel great distances and finally come to rest in a distant basin of deposition. Tephra (pyroclastic debris) ultimately originates from a volcanic source and may accumulate in the same basin of deposition in which the volcanoes are located or the particles composing them may travel long distances, just as terrigenous particles, to accumulate ultimately in a remote basin of deposition. Both are considered pyroclastic rocks, defined as rocks formed from tephra. Table 7-1 is an outline of groups of sedimentary rock categories.

This exercise focuses on the Group I, or intrabasinal, rocks. Like all rocks, they may be classified and named (1) in a purely objective, descriptive manner, or (2) on the basis of what is known, inferred or believed to be the origin of the rocks. Some classifications incorporate both objective and genetic elements in designating a rock as belonging to a particular category or type.

Group	Category
I. Intrabasinal rocks	1. Carbonate rocks: limestone and dolostone 2. Authigenic rocks 3. Carbonaceous rocks
II. Extrabasinal rocks	4. Terrigenous rocks
III. Pyroclastic rocks	5. Pyroclastic rocks

Sedimentary Rock Groups
(Friedman and Sanders, 1978, Table 7-1, p. 170)

Table 7-1

Intrabasinal rocks may be carbonate, authigenic or carbonaceous (Table 7-1) and by definition are composed of particles derived from within the basin of deposition. In terms of volume in the Earth's crust, the carbonate rocks (limestone and dolostone) comprise a large percentage of those sedimentary rocks underlying the continents.

Limestones

Although geologists have studied carbonate rocks from various points of view for the past 150 years, surprisingly the first widely adopted classifications did not appear in the geologic literature until 1958, 1959, and 1962. The interest in limestones of many geologists was centered on faunal remains. Those who studied only the fossils did not bother about the other elements of limestones. In particular, they did not do much about classifying and naming these rocks. In fact, in the mid-1950's, when petroleum companies discovered major reservoirs of petroleum in carbonate rocks, the geological profession, and specifically geologists in oil companies, realized their widespread lack of basic understanding of limestones.

The economic incentive spurred various crash programs in oil companies to develop usable classifications. Throughout much of the 1950s, geologists from several major oil companies worked simultaneously on the problem of classifying and naming limestones. For reasons of competition, communication between these companies did not exist. Coincident with, and partly encouraged by, the new economic interest came a stimulation of similar research in colleges, universities, and various geological surveys. Hence, it is surprising that when in 1962 the different classifications finally appeared in print, the differences between them were only slight. Still, some differences did exist, which reflected in part the area of study in which the geologists did their work or the particular philosophy the geologists adopted in classifying.

Except for reefs, which framework builders construct, or for consolidated lime mud, the basic constituents of most limestones are sand-size particles. The space between these particles is occupied by (1) a matrix of lime mud, whose lithified equivalent is known as micrite, (2) a calcite cement or spar, or (3) void space. Micrite is lithified, mechanically deposited lime mud, which occurs either as a matrix among sand-size particles in a limestone or as the only particles in a fine-grained limestone. In micrite, the particle sizes are $< 4\ \mu$, usually 1 to 3 μ. In a hand sample of a limestone, micrite is dull and opaque; in a thin slice viewed under the petrographic microscope, it is subtranslucent with a faint brownish cast. Sparry cement is the clear, crystalline component of the rocks, having vitreous luster, that occupies the interstices among the sand-size particles; crystals of sparry calcite cement measure tens or hundreds of microns in diameter. The void space (or pores) designates the area between the sand-size particles not occupied by micrite or cement. This void space may be filled with fluids, such as oil or water. The kinds of sand-size particles in a limestone are those discussed in Exercise 1: skeletal particles, oöids, intraclasts, grapestones, and pellets. As the basis for classifying and naming limestones, all schemes of classification employ (1) sand-size particles, (2) micrite, and (3) sparry cement.

The three classifications, which appeared nearly simultaneously, were proposed by M. W. Leighton and C. Pendexter, by R. J. Dunham, and by R. L. Folk.

<u>Leighton-Pendexter Classification</u>. The Leighton-Pendexter limestone classification is the product not only of these two geologists, but of many of their associates of the Standard Oil Company of New Jersey, now Exxon, and its affiliates. The originators of this classification, like those of the other two classifications, recognized that the size-frequency distributions of the particles composing limestone generally are bimodal.

To reemphasize, limestones basically consist of (1) coarse particles, generally of sand-size or larger, and (2) fine particles of silt and clay (when lithified, micrite). According to the Leighton-Pendexter classification, summarized in Table 7-2, besides the organic framework builders such as reefs and such deposits as caliche, travertine, and tufa, limestones are classified and named by their ratios of coarse particles to micrite.

GRAIN /MICRITE RATIO (a)	% GRAINS (b)	GRAIN TYPE (c)					*Organic Frame-Builders*	*No Organic Frame-Builders*
		Detrital Grains	*Skeletal Grains*	*Pellets*	*Lumps*	*Coated Grains*		
		Detrital Ls.	Skeletal Ls.	Pellet Ls.	Lump Ls.	Oolitic Ls. Pisolitic Ls. Algal encr. Ls.	Coralline Ls. Algal Ls. Etc.	
9 : 1	~90%							
		Detrital-Micritic Ls.	Skeletal - Micritic Ls.	Pellet - Micritic Ls.	Lump - Micritic Ls.	Oolitic - (Pisolitic-Etc.) Micritic Ls.	Coralline-Micritic Ls. Algal-Micritic Ls. Etc.	Caliche Travertine Tufa
1 : 1	~50%							
		Micritic - Detrital Ls.	Micritic - Skeletal Ls.	Micritic-Pellet Ls.	Micritic - Lump Ls.	Micritic - Oolitic (Pisolitic Etc.) Ls.	Micritic-Coralline Ls. Micritic - Algal Ls. Etc.	
1 : 9	~10%							
		← Micritic Limestone →						

Leighton-Pendexter Limestone Classification

Table 7-2

The coarse particles include detrital and skeletal particles, pellets, lumps, and coated particles. Leighton and Pendexter used the term detrital for (1) fragments derived from older limestones (i.e., terrigenous sediment) and for (2) intraclasts (Fig. 7-1).

Because they were sediment once before, broken pieces from older limestones were termed second-cycle. Such second-cycle particles in limestones, however, are not common; hence, for most practical purposes, detrital particles according to Leighton and Pendexter's scheme, mean intraclasts. Skeletal particles are whole or broken tests of organisms (Figs. 1-3 and 7-2); pellets may be fecal or peloids or former intraclasts, such as flat pebbles that have been comminuted and rounded to take on the appearance of fecal pellets (Figs. 1-3 and 7-3).

The reason three kinds of pellets are combined into one category is that from a practical point of view, the three appear the same and thus cannot be distinguished. Lumps are composite grains and include grapestones and oncolites (Figs. 7-4 and 7-5). Coated particles are oöids and

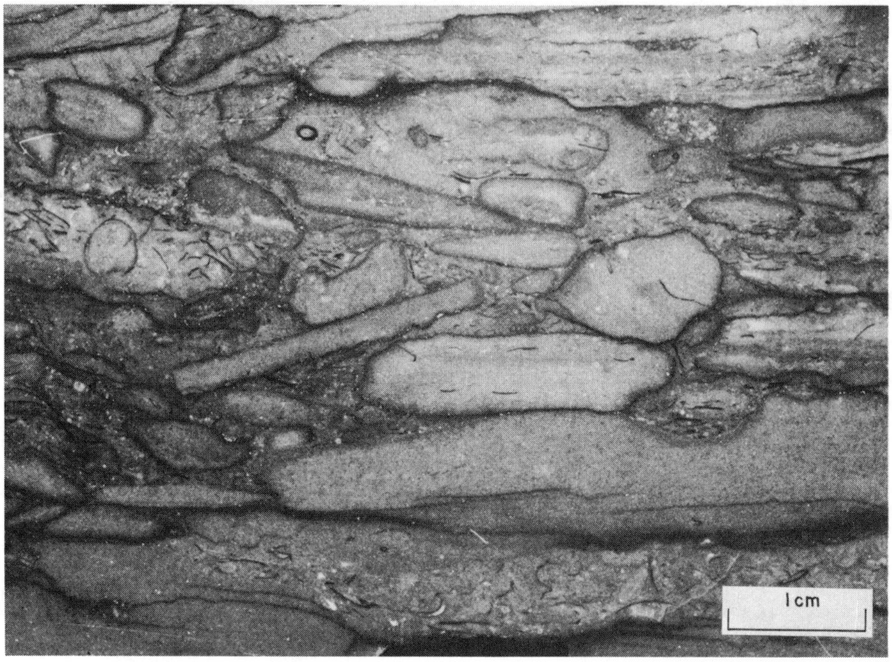

Polished slab of carbonate conglomerate, consisting of intraclasts, cut perpendicular to layers. Each intraclast is an aggregate of lime mud; slablike shapes of cross sections result from breaking along stratification planes (M. Braun, and G. M. Friedman, in Friedman and Sanders, 1978, Fig. 2-31, p. 55).

Figure 7-1

Spines of echinoid. At left, heap of individual spines separated from skeleton. At right, upper surface of skeleton with spines still attached. South shore, Sicily. (Friedman and Sanders, 1978, Fig. 2-18, p. 47.)

Figure 7-2

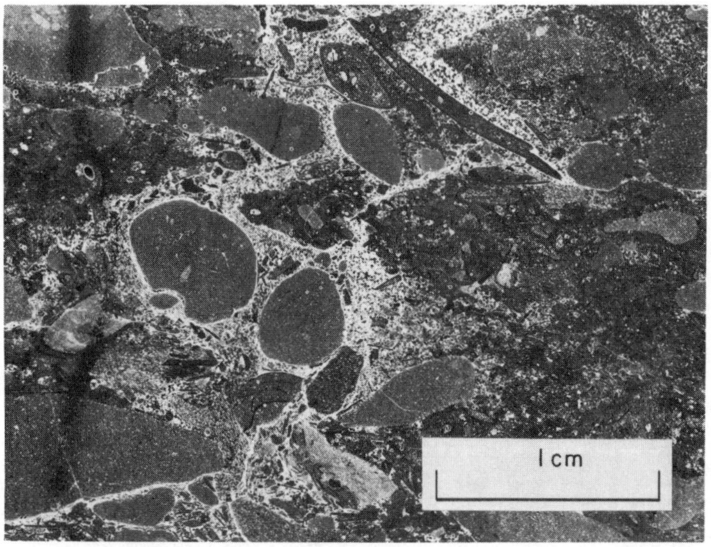

Photograph printed directly from acetate peel of polished and etched surface of limestone containing sand-size and larger aggregates of lime mud. Rounded particles in center resemble fecal pellets. Tribes Hill Formation (Lower Ordovician), Mohawk Valley, New York. (M. Braun, G. M. Friedman, and J. E. Sanders, in Friedman and Sanders, 1978, Fig. 2-25, p. 52.)

Figure 7-3

pisolites (Figs. 1-3 and 7-6) and particles encrusted with foraminifers. Although caliche pisolite, which grows *in situ*, is not strictly a particle (Fig. 7-7), from a practical point of view, it generally cannot be distinguished from coated particles that resemble it.

If more than one kind of particle occurs in a limestone, each may be included in naming the rock. The order should be in terms of relative amounts, with the more-abundant kinds of particles following those that are less abundant, the particle in greatest abundance being mentioned last. For example, a limestone containing 60 percent oöids and 40 percent skeletal fragments is called a skeletal-oölitic limestone. The particle names are separated by hyphens. Adjectives relating to various descriptive aspects of the limestone may be used as modifiers before the rock name, such as particle size "medium-grained, coarse-grained," etc., rock color, or the kind of skeletal material present in the rock, such as crinoidal, molluscan, or other. In fact, the name of the organisms, such as crinoidal or molluscan, may replace the term skeletal.

Limestones consisting of rigid framework builders, such as corals, bryozoa, and algae, are given a separate category in this classification (Table 7-2).

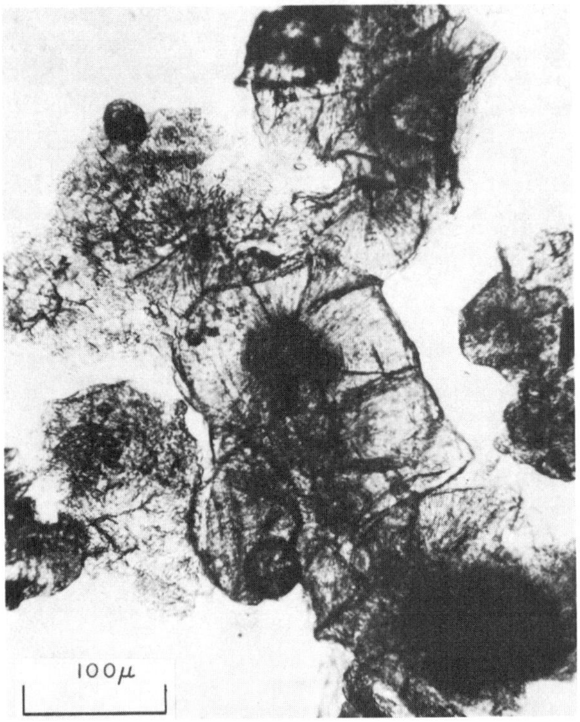

Photomicrograph in plane-polarized light of thin section cut through grapestone clump, algal mats in modern sediments, sea-marginal hypersaline pond, Gulf of Aqaba, Red Sea. Particles are mostly radial oöids. (G. M. Friedman, A. J. Amiel, M. Braun, and D. S. Miller, in Friedman and Sanders, 1978, Fig. 2-30, p. 54.)

Figure 7-4

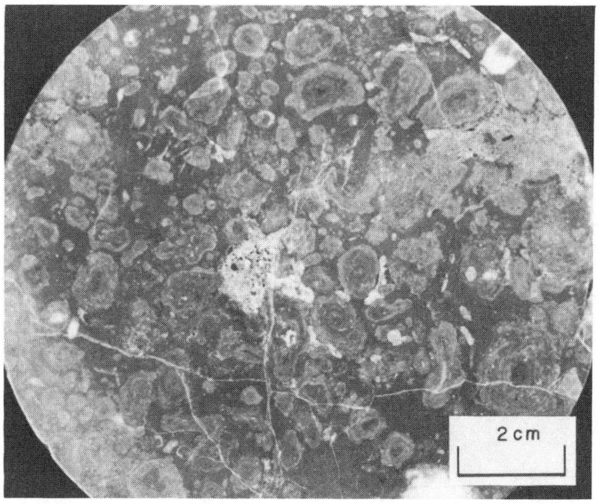

View of oncolites on polished surface, cut parallel to bedding, of core recovered from test boring through limestone of Inmar Formation (Jurassic), Negev Desert, Israel. From this photograph alone, the oncolites cannot be distinguished from vadose pisolites (Fig. 7-7). (M. Goldberg, and G. M. Friedman, 1974, in Friedman and Sanders, 1978, Fig. 2-32, p. 55.)

Figure 7-5

Photomicrograph in plane-polarized light of thin section through oöid having radial fabric. Algal mat, from sea-marginal hypersaline pond, Gulf of Aqaba, Red Sea. (G. M. Friedman, A. J. Amiel, M. Braun, and D. S. Miller, 1973, in Friedman and Sanders, 1978, Fig. 2-28, p. 53.)

Figure 7-6

View of polished surface of dolomitic carbonate rock showing cross sections of vadose pisolites. Compare with Fig. 7-5. Carlsbad Formation (Permian), New Mexico-Texas border. (Friedman and Sanders, 1978, Fig. 2-33, p. 55.)

Figure 7-7

The amount of micrite present should be estimated with a hand lens or in a thin slice under the petrographic microscope. In hand sample, etching of the rock with dilute hydrochloric acid is commonly advisable. The ratio of coarse particles to micrite is an expression of the proportion of larger to smaller particles, and, therefore, is a function of size and sorting. In estimating ratios of coarse particles to micrite, the presence of cement and pore space is disregarded.

Although the Leighton-Pendexter classification, as discussed so far, is descriptive, it has important genetic implications. In Table 7-2, the vertical axis compares the amount of coarse- to amount of fine-grained particles. In the depositional environment, this ratio is controlled by (1) the sorting process, and (2) by local production of coarse particles or lime mud. Ordinarily fine-grained lime mud does not accumulate in areas of continued high-energy waves and currents. Hence, a high ratio of coarse particles to micrite suggests that the water was vigorously agitated and that any available lime mud was removed. The lower the ratio, the less vigorous the agitation of the water. Hence, this ratio can be used as an indicator of the mechanical energy that was expended on the sediment in the depositional environment. The horizontal axis in Table 7-2 gives the kinds of particles or framework builders (disregarding caliche, travertine, or tufa). The kinds of particles that accumulate on various parts of a shelf, such as in the modern Bahamas, reflect the position of a facies in a spectrum of depositional environments. Thus, even the kinds of particles present in a limestone can be used as a clue to the depositional environment.

One of the advantages of the Leighton-Pendexter classification is its use of simple terms that readily convey a mental picture of the limestone. The proposers of this classification did not coin new names.

Dunham classification. The Dunham limestone classification, by R. J. Dunham of the Shell Development Company, distinguishes between limestones consisting of two kinds of rocks in which (1) the original components were bound together, and (2) in which the original components were not bound together. The first group includes limestones showing evidence of binding during deposition (boundstones). In this group, among others, are reefs, stromatolites, and travertine. The second group includes several kinds, to be discussed later, in which the rock consists of coarse particles and micrite. Four classes of limestone of the Dunham classification (Table 7-3), incorporating particles and micrite, compare to the Leighton-Pendexter classification (Table 7-2) with five classes defined by particles on the horizontal axis and three by the particle-micrite ratio on the vertical axis.

CLASSIFICATION ACCORDING TO DEPOSITIONAL TEXTURE

<table>
<tr><td colspan="5">DEPOSITIONAL TEXTURE</td></tr>
<tr><td colspan="4">Original components not bound together during deposition</td><td rowspan="4">Original components were bound together during deposition ... as shown by intergrown skeletal matter, lamination contrary to gravity, or sediment-floored cavities that are roofed over by organic or questionably organic matter and are too large to be interstices.</td></tr>
<tr><td colspan="3">Contains mud
(particles of clay and fine silt size)</td><td rowspan="3">Lacks mud and is grain supported</td></tr>
<tr><td colspan="2">Mud supported</td><td rowspan="2">Grain supported</td></tr>
<tr><td>Less than 10 per cent grains</td><td>More than 10 per cent grains</td></tr>
<tr><td>Mudstone</td><td>Wackestone</td><td>Packstone</td><td>Grainstone</td><td>Boundstone</td></tr>
</table>

Dunham Limestone Classification

Table 7-3

In the Dunham classification, the major criterion for classifying and naming limestones is the abundance of coarse particles and micrite. In this classification, the term lime mud is used instead of micrite; the user of this classification tries to estimate the amount of lime mud present between particles in the original depositional environment. The rationale of this classification is that it allows mapping of gradients in rate of production of sand-size particles relative to rate of accumulation of lime mud. In calm waters, lime mud is able to settle on the bottom and remain there. Hence, rocks that form from an original lime mud deserve to be contrasted with those originally devoid of lime mud, regardless of the number and size of the sand-size particles. This relationship between sand-size particles and lime mud distinguishes an original sediment deposited in calm water from that of a sediment deposited in agitated water. This distinction is fundamental, and in the Dunham classification is incorporated in the class name.

Two categories are known among limestones not bound together during deposition: (1) those that in the depositional environment consisted of sand-size or larger particles which supported one another (particle supported), and (2) those that in the depositional environment consisted of lime mud in which the particles were enclosed or "floated" and did not support one another (mud supported) (Table 7-3). Particle-supported rocks fall into two further classes: (1) particle-supported limestones entirely lacking in lime mud, known as grainstones, and (2) limestones containing lime mud but that are particle-supported, known as packstones. The mud-supported kinds of limestone are subdivided into two additional classes: (1) mud-supported limestones containing <10 percent sand-size or larger particles, known as mudstone, and (2) mud-supported limestones containing >10 percent sand-size or larger particles, known as wackestone (Table 7-3). The critical achievement of the Dunham classification is the distinction between "mud-supported" and "particle-supported." According to Dunham, the names of the classes are tentative; yet the names are short, meaningful English nouns that apply only to the texture of the limestone. These terms can be combined with names describing the kinds of particles present in the rock, such as fusulinid lime packstone, ostracod lime mudstone, or coated-particle lime packstone. This classification inserts the term lime between the name of the class and the modifier referring to the sand-size particle. The term wackestone has been under fire by critics; the other terms have been readily accepted.

Folk classification. The Folk limestone classification follows the pattern of subdividing the constituents of limestones into three groups: (1) particles (meaning sand size or larger), (2) micrite, and (3) cement. Folk designates all particles as allochems; the cement, known as sparry calcite, and micrite are designated as orthochemical constituents or "orthochems." In this classification, which we have slightly modified, four kinds of particles are recognized: (1) intraclasts (including grapestone), (2) oöids, (3) whole skeletons or fragments of fossils, and (4) pellets. The interstices among the four kinds of particles are filled with either sparry calcite cement or with micrite. Therefore, the combinations of sand-size or larger particles with micrite or cement provide a basis for recognizing eight kinds of limestones. In addition, micrite may lack sand-size particles and hence stands alone as a ninth kind of limestone. Finally, a tenth kind is reef rock (Fig. 7-8).

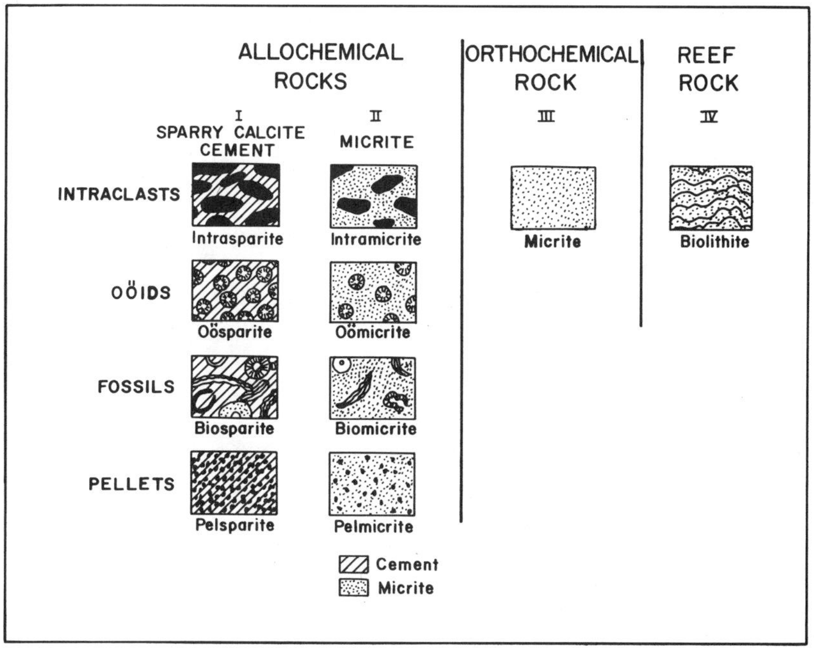

Slightly Modified Folk Limestone Classification
(Friedman and Sanders, 1978, Fig. 7-1, p. 174)

Figure 7-8

These ten kinds of limestone have been assigned names consisting of composite words. Each particle name is abbreviated by the following prefixes:

intraclasts = intra
oöids = oö
fossils = bio
pellets = pel

A limestone having a cement consisting of sparry calcite is known as a sparite. In the Folk scheme, limestones are named by combining the abbreviated particle prefix with either the word sparite or the word micrite. Thus, the names of the basic eight kinds of limestone are intrasparite, intramicrite, oösparite, oömicrite, biosparite, biomicrite, pelsparite, and pelmicrite. Micrite lacking sand-size particles and designated simply as micrite is a ninth name; and reef rock, known as biolithite, is the tenth name of the Folk classification. If several particles are important constituents of a limestone, they are each listed in order of increasing abundance, for instance, intrabiopelmicrite. Where in a sparite some micrite occurs among the particles, such as among oöids, the appropriate name is oömicsparite; conversely, where in a micrite some sparite is present among the particles, the term is oösparmicrite.

This classification allows for particles larger than sand size, which are, however, relatively uncommon in limestones. For particles of pebble size, such as the intraclasts of flat-pebble conglomerates, the shells of lag concentrates, or the pellets of organisms having large anal diameter, the term rudite is employed. A limestone composed of one of those unusually large pellets, if accumulated in a micrite, is known as pelmicrudite; flat pebbles in a micrite would be intramicrudite. If sparry cement occurs between these same pebble-size particles, the limestones would be designated pelsparrudite and intrasparrudite.

The critical contribution of the Folk classification is that this classification forced us to examine what lay between the particles, whether spar or micrite. When this classification was first introduced, many geologists rebelled against it because they felt that the proposed new names were too strange. However, once Folk's classification has been mastered, it proves to be both practical and logical.

Problems and Limitations in Classifying Limestones

As you know from the preceding paragraphs, modern classifications of limestones are based on the relative concentrations of micrite and sparry cement. If these classifications were purely descriptive, this philosophy of using micrite and sparry cement would not raise any difficulties. However, the classifications have genetic implications. They are based on the idea that in calm waters, tiny particles of lime mud are (1) available, and (2) able to settle on the bottom and remain there, whereas in agitated waters, particles of micron size, such as those in lime mud, remain in suspension and are not deposited. Although this concept is satisfactory from the point of view of dynamics, in the depositional environment lime mud does not necessarily remain in place after it is buried. In addition, there are other problems in the study and classification of limestones, such as: (1) the techniques of study employed, (2) the transfer of lime mud, (3) the precipitation of biochemical cement that does not become spar, but rather remains cryptocrystalline, and (4) recrystallization.

Techniques. With a hand lens, especially after a sample has been etched with dilute acid, micrite and sparry cement can usually be distinguished. When this is difficult with a hand lens, examination under a binocular microscope, especially if the etched sample is immersed in water (for instance, a chip immersed in a watch glass), makes the difference between micrite and sparite readily apparent. About 70 to 80 percent of all limestones can be readily classified and named with a hand lens and binocular microscope. For those samples of limestone that defy ready identification, examination may be necessary in a thin slice under the petrographic microscope.

Transfer of Lime Mud. Lime mud from low-energy deposits may filter into empty pores of underlying high-energy deposits, or waters flowing through pores may effect such a transfer of lime mud. Interpretations should not be based entirely on particle-micrite ratios but rather should include consideration of the possibility of such transfer of lime mud from a low-energy to a high-energy deposit.

Precipitation of Cryptocrystalline Cement. An even more complicated problem than the physical transfer of mechanically deposited lime mud is the precipitation of cement that remains cryptocrystalline; that is, under the microscope its crystals are too small to be recognized and separately distinguished and hence this cement looks just like micrite. Such a cement precipitates within millimeters to centimeters of the surfaces of such high-energy deposits as reefs and beachrocks. Since such a cement can be precipitated within the pore spaces of sediments in even the highest of high-energy environments, noncritical use of the particle-micrite ratio as a reliable clue to depositional environment would lead to error.

In summary, the philosophy of using the particle-micrite ratio as a criterion for interpreting the depositional environment of a limestone is based on the common observation that in agitated waters, pores among the particles remain open and later, during diagenesis, may be filled with material that becomes sparry calcite. In many cases this idea is correct. Therefore, the classifications of limestones discussed in this chapter are probably logical and practical. However, one should be aware that such classifications are not infallible, and as in all geological interpretations, caution is necessary.

Recrystallization. The carbonate minerals composing limestones are very susceptible to recrystallization. After these minerals have recrystallized, it may be difficult or impossible to decide, even with a petrographic microscope, what kind of original carbonate sediment was ancestral to the limestone being studied. Because the former particles take on new shapes as crystals, recrystallization is commonly referred to by the name neomorphism (neo = new; morph - shape). We can define neomorphism as a transformation between one mineral and itself or a polymorph, which results in the growth of new crystals that are larger or simply different in shape from the previous particles or crystals. Generally in neomorphism, crystals of calcite replace original particles that were composed either of aragonite or of calcite.

Neomorphism in limestones may proceed in at least three ways: (1) by micrite enlargement, a process by which particles of lime mud that measure only a few microns in diameter enlarge to sizes measuring tens or even hundreds of microns in diameter; (2) by a process in which crystals of calcite, generally of the order of tens to hundreds of microns in size, may replace former micrite, sparite, and particles, without respect for any previously existing boundaries and continuing until the entire limestone consists of these replacing crystals; and (3) by the secondary formation of discrete spherical bodies commonly resembling pellets.

Micrite enlargement creates neomorphic products resembling cement in crystal size. Depending on the crystal size of the new product, the result is known as microspar or pseudospar. Microspar is a mosaic of neomorphic crystals having diameters ranging from 4 to 10 μ or even to 50 μ; pseudospar is a mosaic of neomorphic crystals having diameters larger than those of microspar ($>$10 to 50 μ) (Fig. 7-9). In microspar and pseudospar,

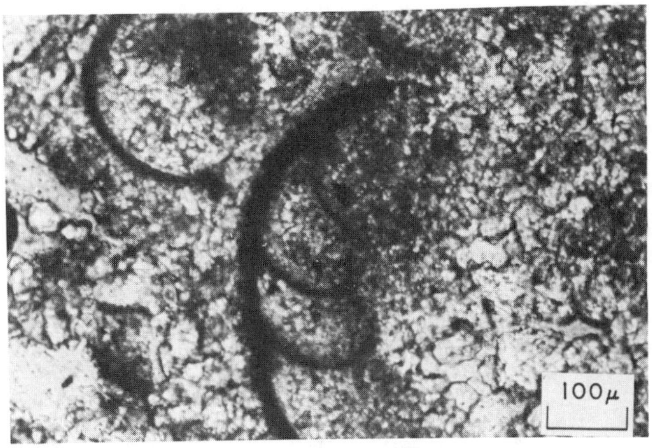

Photomicrograph in plane-polarized light of thin slice of limestone showing neomorphic crystals of calcite that have replaced the former fabric of the rock. Crystals < 50 μ are microspar; those > 50 μ are pseudospar. Ghosts of Foraminifera are preserved as dark lines of carbonate containing opaque inclusions of unknown composition. Pleistocene limestone from borehole, Mediterranean coast of Israel. (L. Ordan, G. M. Friedman, and J. E. Sanders, in Friedman and Sanders, 1978, Fig. 7-2, p. 176.)

Figure 7-9

the distribution of the sizes of individual crystals is usually random and devoid of any regular pattern (Fig. 7-10). Such microspar or pseudospar may exist in isolated patches and thus appears to be "floating" in the midst of micrite. By contrast, true cement is mostly distributed between particles, and the sizes of individual crystals of the cement tend to increase away from the wall of the original void. Even a sample of limestone that has undergone extensive micrite enlargement may display relic patches of nonreplaced micrite, either locally or throughout. These patches reveal that the limestone originally was a micrite rather than a sparite. These features are useful and helpful in naming, classifying, and understanding neomorphically altered limestones.

Crystals of calcite, or as we shall see later also of dolomite or evaporites, on the order of tens or hundreds of microns in diameter, commonly replace particles, cement, and micrite. In so doing these new crystals cut across existing boundaries. Such replacing crystals may be porphyrotopes, which are large crystals surrounded by smaller particles or crystals (Fig. 7-11); or the replacing crystals may be poikilotopes, which are large crystals among smaller ones that enclose original particles or newly formed crystals (Fig. 7-12). Where porphyrotopes or poikilotopes are abundant, especially where they coalesce and merge, the original texture

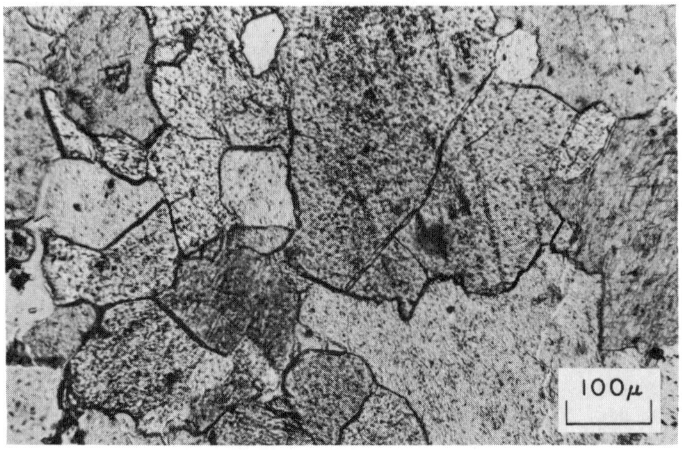

Photomicrograph in plane-polarized light of thin slice of limestone showing neomorphic crystals of calcite that have replaced the former fabric of the rock, including skeletal fragments. Dark parallel lines in large crystal in upper central part of photograph are probably organic matter defining a possible ghost structure of a molluscan shell. Pleistocene limestone from borehole, Mediterranean coast of Israel. (L. Ordan, G. M. Friedman, and J. E. Sanders, in Friedman and Sanders, 1978, Fig. 7-3, p. 176.)

Figure 7-10

Photomicrograph in cross-polarized light of thin slice of dolostone showing crystals of two sizes, (1) a large crystal or porphyrotope, and (2) small crystals enclosing it. Black areas are unfilled pores. Madison Limestone (Mississippian), Lander, Wyoming. (G. M. Friedman, in Friedman and Sanders, 1978, Fig. 7-4, p. 176.)

Figure 7-11

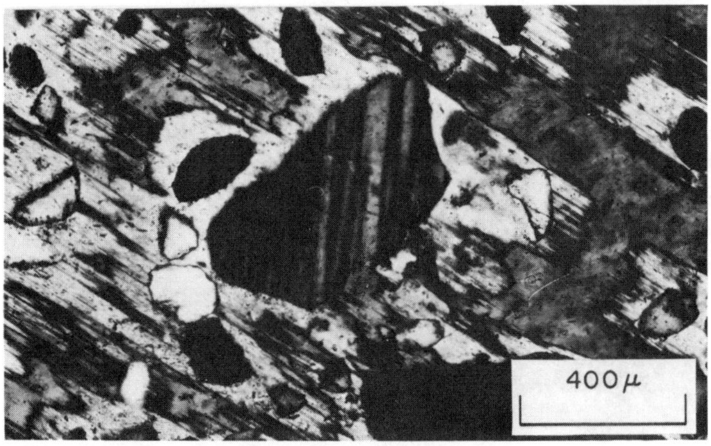

Photomicrograph in cross-polarized light of thin slice showing poikilotopic gypsum (white areas with closely parallel light and dark striae inclined from upper left to lower right) enclosing particles of quartz and feldspar. Holocene. Great Salt Plains near Jet, Oklahoma. (G. M. Friedman, in Friedman and Sanders, 1978, Fig. 7-5, p. 175.)

Figure 7-12

may be entirely obliterated. Fortunately, the original boundaries of particles are commonly preserved as ghosts that are recognizable outlines of the original particles. Within porphyrotopes or poikilotopes, the original pattern of ornamentation of fossils may be preserved. Ghosts preserve original boundaries because organic matter is concentrated along such boundaries. Where proper conditions prevailed, pyrite may have formed along the boundaries of original particles (Fig. 7-13).

Because of the ready susceptibility of carbonate minerals to neomorphic changes in many limestones, even experienced sedimentary petrographers may not always be able to recognize the original particles, cement, and micrite. Neomorphism in many limestones advances to a stage where in their growth all crystals of calcite abut and mutually interfere with one another. In such mutual interference, individual crystals cannot develop their own crystal faces, that is, the outward planar growth surfaces of crystals which reflect their internal structure or lattice. The boundaries between crystals that do not develop true crystal faces are described as anhedral, that is, crystals having boundaries that are not crystal faces. An arrangement of anhedral crystals is known as a xenotopic fabric (Fig. 7-14).

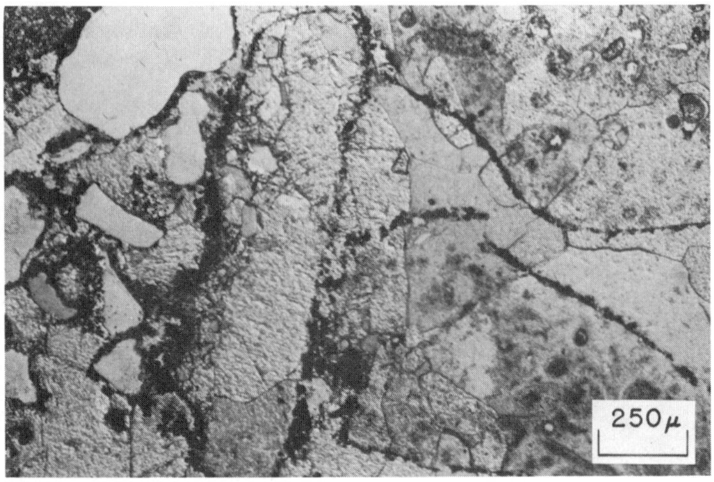

Photomicrograph in plane-polarized light of thin slice of limestone showing ghost of molluscan fragment (approx. center) whose original boundary is preserved by pyrite (black). Pleistocene limestone from borehole, Mediterranean coast of Israel. (L. Ordan, G. M. Friedman, and J. E. Sanders, in Friedman and Sanders, 1978, Fig. 7-6, p. 177.)

Figure 7-13

Photomicrograph in plane-polarized light of thin slice of dolostone showing xenotopic fabric. Cool Creek Formation (Ordovician), near Mill Creek, Oklahoma. (G. M. Friedman, in Friedman and Sanders, 1978, Fig. 7-7, p. 177.)

Figure 7-14

In a xenotopic fabric, the boundaries between crystals that mutually interfered with one another during growth are known as compromise boundaries. To reemphasize, these boundaries are not crystal faces. In limestones displaying a xenotopic fabric, all evidence of former particles, micrite, or cement may have been erased, unless ghosts are preserved that transect the crystals of the xenotopic fabric. Yet, even in less-severely altered limestones, experienced petrographers may be unable to recognize the depositional texture or distinguish cement from pseudospar.

Only careful study of thin sections with a petrographic microscope helps in unraveling the effects of neomorphism and thus in naming and classifying the limestone on the basis of the ancestral calcium-carbonate-sediment. Commonly, even such careful work is inconclusive. In that eventuality, one must be satisfied by naming the rock simply as a recrystallized limestone or as a neomorphically altered limestone.

References Dealing with Limestones

Bathurst, R. G. C., 1971, Carbonate sediments and their diagenesis: Amsterdam, Elsevier Pub. Co., 620 p.

Bramkamp, R. A.; and Powers, R. W., 1958, Classification of Arabian carbonate rocks: Geol. Soc. America, Bull., v. 69, p. 1305-1318.

Dunham, R. J., 1962, Classification of carbonate rocks according to depositional texture, p. 108-121, _in_ Ham, W. E., ed., Classification of carbonate rocks: Tulsa, Okla., Am. Assoc. Petroleum Geologists, Mem. 1, 279 p.

Folk, R. L., 1959, Practical petrographic classification of limestones: Am. Assoc. Petroleum Geologists, Bull., v. 43, p. 1-38.

__________, 1965, Some aspects of recrystallization in ancient limestones, p. 14-48, _in_ Pray, L. C., and Murray, R. C., eds., Dolomitization and limestone diagenesis, a symposium: Tulsa, Okla., Soc. Econ. Paleontologists and Mineralogists, Spec. Pub. No. 13, 180 p.

Friedman, G. M., 1965, Terminology of crystallization textures and fabrics in sedimentary rocks: Jour. Sedimentary Petrology, v. 35, p. 643-655.

Friedman, G. M.; and Sanders, J. E., 1978, Principles of sedimentology: New York, John Wiley & Sons, 792 p., Chapter 7, Classifying and naming sedimentary rocks, with emphasis on historical and genetic aspects, p. 167-194.

Ham, W. E.; and Pray, L. C., 1962, Modern concepts and classifications of carbonate rocks, p. 2-19, _in_ Ham, W. E., ed., Classification of carbonate rocks: Tulsa, Okla., Am. Assoc. Petroleum Geologists, Mem. 1, 279 p.

Leighton, W. M.; and Pendexter, C., 1962, Carbonate rock types, p. 33-61, _in_ Ham, W. E., ed., 1962, Classification of carbonate rocks: Tulsa, Okla., Am. Assoc. Petroleum Geologists, Mem. 1, 279 p.

Wilson, J. L., 1969, Microfacies and sedimentary structures in "deeper water" lime mudstones, p. 4-19, _in_ Friedman, G. M., ed., Depositional environments in carbonate rocks, symposium: Tulsa, Okla., Soc. Econ. Paleontologists and Mineralogists, Spec. Pub. No. 14, 209 p.

Dolostones

Compositionally dolostones are of two kinds: (1) calcareous dolostones, carbonate rocks containing 50 to 90 percent of the mineral dolomite, and (2) those designated simply as dolostone, carbonate rocks containing 90 percent or more of the mineral dolomite.

Dolostones may or may not preserve the original textures of their antecedent limestones. Where dolostones preserve such textures, the outlines of original calcareous particles, now dolomite, such as skeletal fragments, oöids, or pellets, may be readily apparent or may be obscured but still recognizable as ghosts. In such a dolostone, replacement of the preceding limestone must have proceeded parallel to the boundaries or following the boundaries of the original particles, thus preserving their shapes. Such a replacement is known as selective replacement, as the particles are selectively replaced. Dolomite replacing limestone may obliterate all preexisting textures; such replacement is known as pervasive replacement. The product of pervasive replacement is generally featureless.

In dolostones, the distribution of the sizes of the crystals may be bimodal, polymodal, or unimodal. The fabrics denoting such distribution are termed inequigranular or equigranular. An inequigranular fabric is a fabric in which crystals of at least two distinct sizes exist. Usually the sizes differ by orders of magnitude; the size-frequency distribution of the crystals is thus bi- or polymodal. The fabric of inequigranular dolostone may be porphyrotopic or poikilotopic. An equigranular fabric is a fabric in which the crystals tend to be of more or less uniform size.

Under the heading of recrystallization in limestones you were introduced to the terms anhedral and xenotopic. The opposite of anhedral is euhedral, which implies that a crystal is wholly bounded by crystal faces. A fabric in which the predominant constituents are euhedral crystals is known as idiotopic (Fig. 7-15).

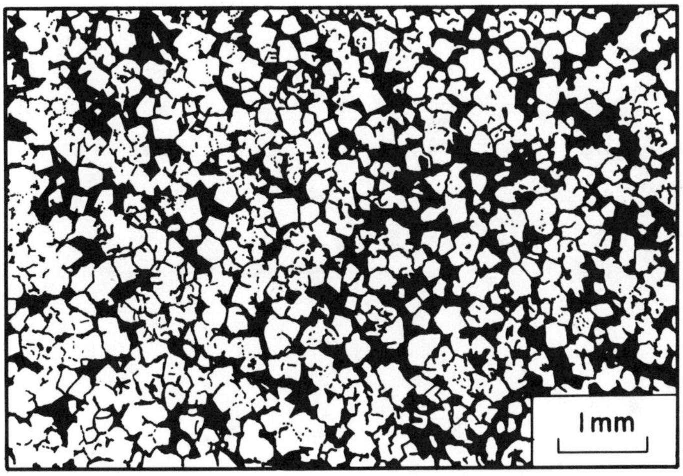

Sketch of view of thin slice of dolostone viewed under polarizing microscope showing a largely idiotopic fabric consisting of noninterfingering euhedral crystals. Dark areas between crystals are open pores. Lower Permian, Tunisia. (G. M. Friedman, in Friedman and Sanders, 1978, Fig. 7-8, p. 178.)

Figure 7-15

Idiotopic dolostone is probably one of the most important carbonate reservoirs for oil and gas. Because of their "sugary" appearance, idiotopic dolostones are commonly known as sucrosic dolostones. Idiotopic dolostones are thought to form as follows: euhedral crystals of dolomite nucleate in a calcium-carbonate host; as they grow, they replace progressively more of their host. Dolomite and calcite have different solubilities in freshwater, calcite is more soluble. After a framework of euhedral dolomite crystals has grown, the remaining nonreplaced calcite is removed by dissolution in freshwater. This leaves a highly porous idiotopic fabric of euhedral dolomite crystals. An idiotopic fabric without porosity cannot possibly develop unless it occurs in a limestone matrix. In a fabric devoid of pores, growing crystals will interfere with one another and euhedral crystal faces cannot develop; but growth at the expense of a host limestone allows new crystals, such as those of dolomite, to develop their euhedral faces within the calcite host. In a fabric lacking porosity, not more than a very few crystals can be euhedral. For an explanation, consider crystals A and B in Figure 7-16 and the junction between them (thickened in the figure). This junction can be one of three kinds. It may be a crystal face of A and contribute to A's euhedralness, but not to B's; it may be a crystal face of B and contribute to B's euhedralness, but not to A's; or it may be a compromise boundary, not a face of either crystal. It cannot be a crystal face of both A and B. Remember that a crystal face is a very specific kind of outward planar surface, namely one that reflects the growth and internal structure or lattice of the crystal. It is not just any planar surface.

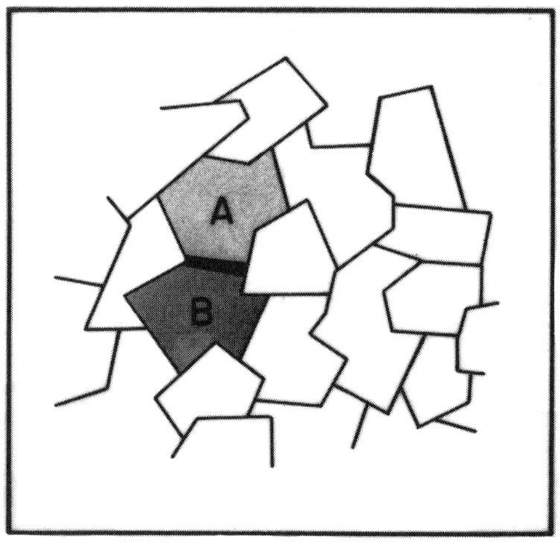

Sketch of crystalline fabric in sedimentary rock. The junctions between crystals can be of three kinds, as illustrated by the junction (thick black line) between crystal A and crystal B (shaded). (1) The junction may be a crystal face of A and contribute to A's euhedralness, but not to B's. (2) The junction may be a crystal face of B that contributes to B's euhedralness, but not to A's. (3) The junction may be a compromise boundary and not a face of either crystal. (R. G. C. Bathurst, _in_ Friedman and Sanders, 1978, Fig. 7-9, p. 179.)

Figure 7-16

Crystals partly bounded by crystal faces are known as subhedral. A fabric in which most of the constituent crystals are subhedral is hypidiotopic (Fig. 7-17).

By continued growth of dolomite crystals, idiotopic dolostone may be converted into hypidiotopic or xenotopic dolostone. This causes the dolomite crystals to grow together and to fill the pore spaces; in such a growth, the crystals progressively interfere with one another.

Both equigranular and inequigranular fabrics may be xenotopic, hypidiotopic, and idiotopic.

Photomicrograph in plane-polarized light of thin slice of dolostone showing a hypidiotopic fabric. Madison Limestone (Mississippian), Dubois, Wyoming. (G. M. Friedman, in Friedman and Sanders, 1978, Fig. 7-10, p. 179.)

Figure 7-17

Descriptive adjectives giving the size of dolomite or other crystals may be added to the name as modifiers. Such modifiers for crystal size include:

micron-sized:	0 to 10 μ (0.001 to 0.01 mm)
decimicron-sized:	10 to 100 μ (0.010 to 0.100 mm)
centimicron-sized:	100 to 1000 μ (0.100 to 1.000 mm)
millimeter-sized:	1 to 10 mm
centimeter-sized:	10 to 100 mm

A further breakdown of the centimicron-size range may be made by using the terms fine crystalline (100 to 250 μ), medium crystalline (250 to 500 μ), and coarse crystalline (500 to 1000 μ). This further breakdown into three additional size classes is based on empirical observations of the distribution of crystal sizes in dolostones and recrystallized limestones and is therefore useful; it does not, however, follow the metric framework of the above-given size classification.

The naming and classifying of these various fabrics and size classes can also be extended to limestones that have undergone neomorphism or to precipitates, such as evaporites.

Dolostones may also be classified genetically into three major groups: (1) syngenetic dolostone, (2) diagenetic dolostone, and (3) epigenetic dolostone. Syngenetic dolostone is here defined as dolostone that has formed penecontemporaneously in its environment of deposition. This kind of dolostone contrasts with diagenetic dolostone, which is dolostone formed by replacement of calcium-carbonate sediments or of limestones during or following consolidation, such as within beds of carbonate sediments or limestones. In borderline cases, the distinction between syngenetic and diagenetic becomes difficult or impossible to make. Epigenetic dolostone is here defined as dolostone that has formed by localized replacement of limestone along postdepositional structural elements, such as faults and fractures. Many, but not all, epigenetic dolostones are genetically associated with metallic ore deposits, notably of lead and zinc minerals.

References Dealing with Dolostones

Bathurst, R. G. C., 1975, Carbonate sediments and their diagenesis, 2nd ed.: Amsterdam, Elsevier Pub. Co., 658 p.

Dunbar, C. O.; and Rodgers, John, 1957, Principles of stratigraphy: New York, John Wiley & Sons, 356 p.

Friedman, G. M., and Sanders, J. E., 1967, Origin and occurrence of dolostones, p. 267-348, *in* Chilingar, G. V.; Bissell, H. J.; and Fairbridge, R. W., eds., Carbonate rocks, origin, occurrence, and classification: Amsterdam, Elsevier Pub. Co., 471 p.

Shrock, R. R., 1948, A classification of sedimentary rocks: Jour. Geology, v. 56, p. 118-129.

Zenger, D. H.; Dunham, J. E.; and Ethington, R. L., eds., 1980, Concepts and models of dolomitization: Soc. Economic Paleontologists and Mineralogists, Spec. Publ. No. 28, 320 p.

Authigenic Rocks

Authigenic rocks are chemically and mineralogically quite diverse; they have in common only their authigenic origin. The most important kinds of authigenic rocks are: (1) chert, (2) evaporites, (3) phosphate rock, (4) sedimentary iron ores, and (5) manganese rock.

Chert. Chert is a tough brittle rock consisting of nearly pure silica, exhibiting a splintery to conchoidal fracture and a vitreous luster. Common synonyms for chert or varieties of chert include jasper, flint, novaculite, and porcellanite. Jasper is yellow-, red-, brown-, or green-colored chert. Flint is gray to black, cryptocrystalline chert that exhibits conchoidal fracture and whose color derives from included organic matter; novaculite is a chert that in hand specimen is a milk-white color; and porcellanite is a low-density and dull-lustered rock having the texture and appearance of nonglazed porcelain. In thin slice under the petrographic microscope, chert may be cryptocrystalline, fine crystalline, medium crystalline, coarse crystalline, or fibrous. The fibrous variety is chalcedony, a polymorph of quartz, which displays a distinctive pattern of radial fibers; the others are quartz. Porcellanite consists of cristobalite or opal; cristobalite is a metastable polymorph of quartz; opal is a hydrous silica that contains between 4 and 9 percent water. This water is physically absorbed; most of it is driven off at 100°C. The cristobalite of porcellanite is highly disordered and is known as opal-CT. In the Monterey Formation (Miocene) of California, porcellanite represents a transitional phase between opal and quartz.

Chert occurs (1) in nodules, and (2) in strata. The mineral composition of chert appears to be influenced by the composition of the host sediment or host rock. As an example, in the central Pacific Ocean, chert composed of cristobalite is found in clay-rich sediments, whereas quartzose chert is restricted to carbonate sediments.

On the basis of crystal size, the quartz of chert consists of (1) microquartz and (2) megaquartz (Fig. 7-18). Microquartz consists of quartz crystals $< 35\ \mu$, as measured in thin sections; megaquartz consists of quartz crystals $> 35\ \mu$ in thin sections.

Tests of siliceous organisms, such as those of radiolarians and diatoms as well as the spicules of sponges, consist of opal. Likewise, siliceous shells in cherts of Tertiary age are mostly opaline, but nearly all silica in Paleozoic cherts is present as quartz and chalcedony. Other silica minerals that occur in chert include quartzine and lutecine, which are fibrous varieties of quartz and lussatite, respectively. The mineral lussatite originally was considered to be a fibrous variety of cristobalite, but recently has been identified as a highly disordered cristobalite

Photomicrograph in cross-polarized light of thin slice of chert showing crystals of megaquartz (bulk of photograph) filling cavity lined by a zone of fibrous chalcedony. Long axes of chalcedony fibers stand normal to wall of cavity. Fort Ann Formation (Lower Ordovician), New York State. (S. J. Mazzulo, in Friedman and Sanders, 1978, Fig. 7-11, p. 180.)

Figure 7-18

(opal-CT). Quartzine and lutecine may form at the expense of sulfate minerals in evaporite deposits. The presence of these silica minerals may be the only testimony to the former presence of such sulfates, now replaced and vanished. These silica minerals may therefore be useful environmental indicators.

The origin of chert has been under debate since the beginning of this century, inasmuch as diagenetic processes have generally obliterated all preexisting textural features and faunal remains. The silica of chert may be (1) an alteration product of volcanic rock, such as montmorillonite and volcanic glass, and (2) a precipitate derived from the dissolution of tests of siliceous organisms. The dissolution of the tests of radiolarians is the most probable source of silica in the cherts of the central Pacific Ocean. The silica of most oceanic cherts probably is of biogenic origin. Biogenic opal is dissolved and reprecipitated as finely crystalline cristobalite or opal-CT which inverts to quartz. The end product is a classic dense, vitreous chert. However, although most cherts are probably the end product of the reaction biogenous opal→cristobalite or opal-CT→quartz, the possibility of direct crystallization of quartz should not be excluded. Cherts are not forming in modern sediments.

Impurities in chert generally constitute $<$ 10 percent. Both replaced and nonreplaced fossils are common, as are disseminated detrital quartz, rhombs of carbonate minerals, pyrite, hematite, organic carbon, and clay minerals.

Evaporites. Evaporites are authigenic rocks whose name derives from their origin; they form by precipitation from brines that concentrate by evaporation. Although almost 40 different precipitate minerals have been recorded from evaporite deposits, only about 20 are present in more than trace amounts. Of these only two kinds are common, (1) sulfates, and (2) halides, both of which form extensive deposits of sedimentary rock.

Two kinds of sulfate minerals that form sedimentary rocks are common, (a) gypsum ($CaSO_4 \cdot 2H_2O$), and (b) anhydrite ($CaSO_4$). Although anhydrite forms in modern sea-marginal flats of Abu Dhabi in the Persian Gulf, for all practical purposes, gypsum, as the hydrated sulfate, occurs on the surface, and anhydrite, the anhydrous sulfate, occurs in the subsurface. The change from one kind of sulfate to the other depends on the geothermal gradient as well as on the salinity of the subsurface brine. Anhydrite generally will replace gypsum at depths of 300 to 700 m (Fig. 7-19). Depending on availability of water, gypsum after anhydrite will appear at varying depths, normally between 100 and 300 m.

Both anhydrite and gypsum rocks may be nodular, massive, or laminated (Fig. 7-20).

Gypsum rock may display graded beds, cross strata, scour-and-fill structures, ripples, and other features of current activity. Gypsum rock consisting of selenite crystals may grow in rosette or radial patterns, with some crystals 2 to 2½ m in length. In the subsurface, anhydrite is usually compact and devoid of porosity. Thus such anhydrite acts as a caprock to reservoirs of oil and gas.

Among the halides, halite rock (NaCl; rock salt) is the most common; it occurs in successions up to 1000 m thick. Halite rock is commonly laminated; dark, inclusion-rich layers alternating with light, inclusion-poor layers. Inclusions consist predominantly of anhydrite. Although they are much less abundant than anhydrite, other impurities include dolomite, quartz, calcite, and various clay minerals. The clay mineral talc in evaporites, especially in halite rock, is of authigenic origin. Freshly broken samples of halite rock commonly emit a strong odor of H_2S, suggesting that much sulfate has been reduced. Nonrecrystallized and nondeformed halite rock may show various sedimentary structures, including those of stromatolites and nodules. Such halite rock has formed at the expense of

Acid-etched sample of core from borehole showing laths of anhydrite pseudomorphs after gypsum in a limestone matrix. The anhydrite has the habit of gypsum proving the original sulfate mineral was gypsum; in the subsurface the gypsum changed to anhydrite. Salina Group (Silurian), Michigan. (R. D. Nurmi, in Friedman and Sanders, 1978, Fig. 7-12, p. 181.)

Figure 7-19

organic matter, carbonate, and anhydrite. Anhydrite and dolomite are commonly interlaminated with the halite.

Halite rock commonly recrystallizes and moves as a solid; this movement is referred to as salt flowage. Such recrystallized halite is composed of centimeter-sized crystals having a tightly interpenetrating framework. During recrystallization, inclusions are expelled from halite crystals; these former inclusions are commonly found along the boundaries of crystals rather than within the crystals.

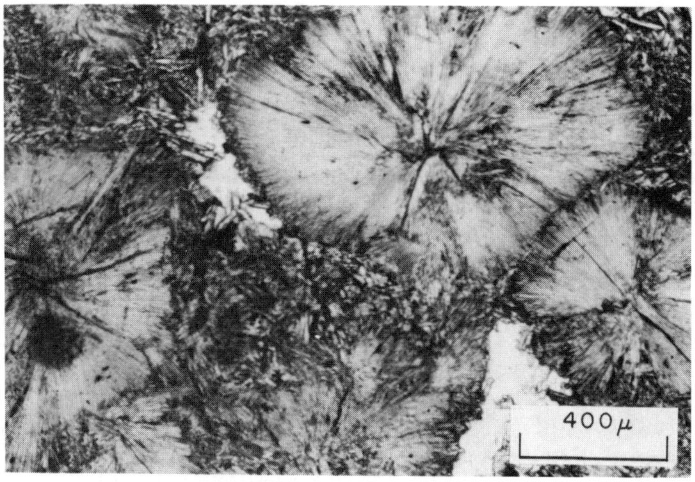

Photomicrograph in plane-polarized light of thin slice of anhydrite rock showing anhydrite nodules with stellate arrangement of crystals; small anhydrite laths are interspersed between the nodules. From Miocene strata encountered in borehole below floor of western Mediterranean Sea (site 124, Deep Sea Drilling Project). (G. M. Friedman, in Friedman and Sanders, 1978, Fig. 7-12, p. 181.)

Figure 7-20

Phosphate Rock. Phosphates or phosphorites (the two terms are synonymous), here referred to as phosphate rock, are composed of the mineral apatite. This rock-building mineral consists of three kinds, which form an isomorphous series: (1) fluorapatite [$Ca_5(PO_4)_3F$], (2) chlorapatite [$Ca_5(PO_4)_3Cl$], and (3) hydroxyapatite [$Ca_5(PO_4)_3OH$]. Phosphate rock of which the essential mineral component is usually a carbonate fluorapatite [$Ca_5(PO_4)_3(F, CO_3)$], a variety of apatite known as francolite, is the basic raw material for many phosphorous-containing compounds on which modern technological developments depend. When the apatitelike phase cannot be identified, the name collophane is commonly applied. Ca, P (or PO_4), and F may be substituted by other elements. Some of these substitutions are potentially of economic interest, because many trace elements are present in higher-than-normal concentrations. Perhaps at some time in the future, uranium, rare earths, and other trace elements may be extracted from phosphate rocks.

Phosphate rock occurs on the modern sea bottom and in ancient deposits. Phosphate-rich sediments are present on the modern continental shelves and as rock nodules on the modern sea bottom extending from shallow to great water depths. Such nodules occur off the coast of California. In some nodules, the concentration of phosphate reaches 95 percent, but in others, abundant impurities of various marine products are included, such as organic matter, siliceous tests of organisms, calcareous shells, some of which are partly phosphatized, sharks' teeth, and various detrital particles. Some nodules contain iron-rich or manganese-rich coatings and appear to be glazed or polished. The internal structures of the nodules are variable; some are homogeneous, others are concentrically banded.

Ancient deposits of phosphate rock occur in extensive phosphogenic provinces as bone fragments, sand-sized pellets of possible fecal origin (100 to 400 μ in diameter), oöids (Fig. 7-21), pisolites, phosphatized tests of various calcareous shells, and large pellets (2 to 5 mm), which have been termed coprolites.

Photomicrograph in plane-polarized light of thin slice of phosphate rock showing oöids composed of collophane. Phosphoria Formation (Permian), Idaho. (P. J. Cook, in Friedman and Sanders, 1978, Fig. 7-15, p. 182.)

Figure 7-21

The term coprolite defines particles of animal feces, mostly of invertebrates, commonly 1 to 15 cm in length, which are light to dark brown or black in color, and usually of elongate to ovoid shape. Some phosphate rock displays cross strata. The average phosphate rock of the Phosphoria Formation contains 80 percent apatite; impurities include quartz, illite, organic matter, dolomite, calcite, and iron oxides.

Sedimentary-Iron Ores. Sedimentary iron ores are iron-rich sedimentary rocks. These rocks have influenced modern histroy.

Although all sedimentary rocks contain readily detectable amounts of iron, the term iron-rich deposits designates those consisting predominantly of iron minerals, notably oxides, hydroxides, carbonates, silicates, and sulfides. The oxides are hematite and magnetite; the hydroxide mineral is goethite (including limonite); the carbonate mineral is siderite; the silicate minerals are chamosite, greenalite, and glauconite; and the sulfide minerals are pyrite and pyrrhotite.

One of the remarkable features of sedimentary iron ores is the contrast in mineral composition between those of Precambrian and those of younger ages. In Precambrian deposits, hematite may occur interbedded with chert, whereas in Phanerozoic deposits, hematite occurs as a replacement of oöids, fossil fragments, or as an earthy matrix. Goethite occurs mostly as a replacement of oöids. Siderite occurs as a sparry cement or as microspar or as a replacement of chamosite or of limestones in which it has replaced various kinds of calcareous particles and matrix or cement, but again notably oöids. Siderite also forms layers or discontinuous concretions. Such layers and concretions, which are in fact siderite rock, are commonly referred to as clay ironstones. Chamosite, the most-common Phanerozoic iron-silicate mineral, occurs mostly as chamosite mudstone or as a constituent of oöids and less commonly as scattered flakes in the matrices of oölitic deposits. Greenalite is almost entirely restricted to Precambrian rocks; in these it occurs as small dark-green granules. In this discussion, we describe only authigenic rocks; hence, such deposits as magnetite or ilmenite sands or sandstones are excluded.

Sedimentary iron ores are classified into three broad groups: (1) bog-iron deposits, (2) ironstones, and (3) iron formations. The basis of the classification combines mineral composition, occurrence, and geologic age. In addition, sedimentary iron ores may be named according to their dominant iron mineral, thus siderite rock, hematite rock, or magnetite rock.

Swamps and lakes of glaciated areas are the sites of bog-iron deposits. Goethite is the iron mineral that composes these deposits. The particles of such deposits may be oölitic and are cemented to form lenses or layers.

Ironstones are ferruginous rocks that are (1) younger than Precambrian in age, and (2) essentially devoid of chert. In these two respects ironstones differ from iron formations. Ironstones are interbedded with limestones, shales, sandstones, and conglomerates. Some consist of oöids and fragments of bryozoa, crinoids, trilobites, and various other skeletal debris partly coated or replaced by iron oxide. Others consist entirely of oöids of hematite in a calcite matrix, or of small, flattened concretions of hematite with fragments of fossils replaced by hematite, and still others are oölitic limonite-chamosite-siderite rocks.

Iron formations are of Precambrian age. Typically, these ferruginous rocks are predominantly interbedded with quartz or chert. Such well-developed bedding has led to the name "banded-iron formation" for those deposits (Fig. 7-22). Hematite, magnetite, and less commonly siderite are the constituent minerals of iron formations.

Photograph of exposure of "banded-iron formation" showing interbedded magnetite (dark gray to black) and quartz (light gray to white). Precambrian, Algoma District, Ontario, Canada. (Friedman and Sanders, 1978, Fig. 7-17, p. 185.)

Figure 7-22

Carbonaceous Rocks. Carbonaceous rocks, mostly the rock coal, have been classified in two ways: (1) by rank (content of carbon and calorific value), and (2) by petrographic characteristics.

In order of increasing rank, the classes of coal are: (1) lignite, (2) subbituminous coal, (3) bituminous coal, and (4) anthracite. The progression of carbonaceous material through the continuous series, lignite (or its precursor peat) through bituminous coal to anthracite, is known as coalification. During coalification, the products darken in color and increase in luster. As they do so, their relative contents of carbon and calorific value gradually rise, and their contents of moisture, volatile matter, and oxygen decrease. Except in anthracite (in which hydrogen decreases), the proportion of hydrogen remains roughly constant through this series. Coalification is accomplished in two stages. The first stage is biochemical; it operates during bacterial conversion of plant material to peat. The second stage is geochemical; during it both chemical and physical changes occur. These are related to pressure, shearing stresses, and temperature.

In essence, rank of coal is a measure of diagenetic and metamorphic grade that reflects the pressure, shear, and temperature to which the coal has been subjected. In closely folded strata of Pennsylvanian age in Pennsylvania, anthracite is present. By contrast, in gently folded strata of the same age, bituminous coal occurs. Tertiary coals in Alaska that have been folded are of high rank, whereas coals of the same age in nonfolded strata in the Dakotas are still lignite. Thus, shear induced by folding can increase the rank of coal. Likewise, older coals are more likely to have been more deeply buried, and hence exposed to greater pressure and shear than younger coals. Thus, Carboniferous coals are generally of higher rank than Tertiary coals. Temperature is also important. The more-intensely folded parts of coal-bearing strata are not necessarily of higher rank than coals in nonfolded strata. Where the geothermal gradient is low, coals may be subbituminous; where the gradient is high, in contrast, the coal may be bituminous at a comparable depth. Thermal metamorphism near igneous intrusives likewise leads to increasing rank.

Using petrographic characteristics, we can recognize four kinds of coal: (1) vitrain (or anthraxylon), a bright, glossy, vitreous or glassy-looking coal having conchoidal fracture; (2) durain, a dull coal, lacking luster and having an earthy appearance; (3) clarain, a coal having a smooth surface when broken at right angles to the bedding plane, such broken faces having a pronounced gloss or shine and silky luster; and (4) fusain (or "mineral charcoal" or "mother of coal"), carbonized wood resembling charcoal; fusain is the constituent of coal that soils the hands when touched. These four kinds of coal can be recognized in hand samples, but are more readily studied by examination of polished surfaces under the microscope using reflected light with oil-immersion objectives. Although we have listed them as individual entities, these four kinds of coal commonly are interlaminated.

Most coals form from accumulated plant debris in sea-marginal swamps. In position with respect to the shoreline, most coals are analogous to anhydrite rock accumulating in sea-marginal flats. The difference between anhydrite and coal as a product in analogous depositional environments is one of climate. Whereas anhydrite forms under arid conditions, coal forms under humid conditions. The warmer and wetter the climate, the more luxuriant is the flora and the greater the accumulation of plants to yield coal. Coals are associated with cyclical sequences of strata that reflect rapid transgressions and regressions and that may have involved deltas.

References Dealing with Authigenic Rocks

Bentor, Y. K. (ed.), 1980, Marine phosphorites - geochemistry, occurrence, genesis: Tulsa, Oklahoma, Soc. Economic Paleontologists and Mineralogists, Special Publ. No. 29, 249 p.

McBride, E. F.; and Thomson, A., 1970, The Caballos Novaculite, Marathon Region, Texas: Geol. Soc. America, Spec. Paper 122, 129 p.

Stanton, R. L., 1972, Ore petrology: New York, McGraw-Hill Book Co., 713 p.

Teichmüller, M.; and Teichmüller, R., 1967, Diagenesis of coal (coalification), p. 391-415, *in* Larsen, G.; and Chilingar, G. V., eds., Diagenesis in sediments: New York, Elsevier Pub. Co., 551 p.

Procedure

Using the specimens of intrabasinal rocks provided for study, carefully determine the compositional and textural character of each one and record your observations on Form 7-1. Examination should include use of hand-lens, dilute HCl, binocular microscope and petrographic microscope, as appropriate.

After recording observational data, name each specimen. For those specimens which are limestones, name them according to the Leighton-Pendexter Classification, Dunham Classification, and Folk Classification. For those specimens which are recrystalized limestones or dolostones, indicate the crystallization texture and fabric using the terminology outlined on pages 132 through 137 and 138 through 141.

REPORT ON CLASSIFICATION OF
INTRABASINAL SEDIMENTARY ROCKS

Exercise 7
Name ____________________
Laboratory Section ____________________
Date ____________________

Name	
Texture	
Composition	
Speci-men No.	

Form 7-1

Speci- men No.	Composition	Texture	Name

Form 7-1 (continued)

When we pretend to limit the operations of nature, to suit our contracted ideas, we most probably do her injustice.

William Maclure
1818

EXERCISE 8

DESCRIPTION AND CLASSIFICATION OF SEDIMENTARY ROCKS IN HAND SPECIMEN AND THIN SECTION: EXTRABASINAL AND PYROCLASTIC ROCKS

Introduction

The debris created by the wearing down of the land includes particles ranging through a wide spectrum of sizes from micron size - even from fractions of a micron - to boulders measuring meters across. These particles are the materials of which extrabasinal rocks are made. Rivers carry particles in mud-, sand-, and even boulder-size ranges. The rivers deliver their loads to estuaries or deltas and from there the particles may move deeper onto the floor of a basin of deposition where they finally come to rest. The sources of the particles thus lie outside the basin of deposition in which the particles finally accumulate, hence the term extrabasinal (see Table 7-1). This term is in contrast to carbonates, authigenic and carbonaceous rocks, which form and remain within the basin of deposition and hence are known as intrabasinal.

Extrabasinal rocks are generally known as terrigenous rocks, a term taken from the Latin word terra, meaning land. These rocks are derived as sediments from weathering and erosion of land masses, hence this name. Pyroclastic rocks are lithified tephra. Their particles originate as explosive, igneous material, but are deposited as sediments.

The purpose of this exercise is to introduce you to the characteristics and nomenclature of the extrabasinal (terrigenous) rocks and pyroclastic rocks.

A fundamental basis for the classification of terrigenous rocks is size. On the basis of particle size, we recognize three main kinds of terrigenous rocks: (1) shales, (2) sandstones, and (3) conglomerates and sedimentary breccias. Grabau (1913) introduced the terms lutite for shale, arenite for sandstone and rudite for conglomerates. A fourth kind of terrigenous rocks, siltstone, is sometimes recognized. For it, the term siltite has been introduced. (The addition of ite from the Greek to an ordinary English word "silt" is a true example of a Greek barbarism.) We shall discuss how to classify and name each of these three kinds of rock separately; a discussion of the fourth kind, siltstone, will be included with that of shale.

Shales

Naming and classifying fine-grained terrigenous rocks is complicated. Various bases for naming these rocks have been proposed, including (1) particle size, (2) proportion of clay minerals, and (3) fissility, the property of splitting easily into thin layers parallel to the bedding.

Probably the most satisfactory definition of shale is according to particle size. In terms of size, shale is the lithified equivalent of mud. Mud is a sediment consisting of clay- and silt-size particles. Likewise, shales are terrigenous rocks composed of clay- and silt-size particles. Surprisingly, despite the technique of breaking down a shale into its constituent particles using immersion of a sample in a bath of an ultrasonic transducer, few reliable data exist on the size-frequency distribution in shales. Simple hand-specimen inspection indicates that many rocks named shales by geologists are really siltstones; these rocks consist of particles of silt rather than of clay size. Despite the use of the term shale for fine-grained sedimentary rocks spanning both the clay- and silt-size ranges, it is important to distinguish the dominantly clay shales from the dominantly silt shales. This difference is very significant in recognizing the transition between bottomsets and foresets in sediments of fine-grained deltas. Some geologists like to make a distinction between shales and siltstones; others are satisfied to lump under the term shale all fine-grained terrigenous rocks.

The proportion of clay minerals in shales varies widely. The commonly accepted notion that shales are composed almost entirely of clay minerals is generally incorrect. In most shales quartz may compose between one-quarter and one-half of the total; in some shales, quartz together with feldspar are even more abundant than one-half. Increasing proportions of quartz and feldspar generally parallel increasing particle size; thus coarse-grained shales generally contain a greater proportion of quartz and feldspar, whereas fine-grained shales have less of these and a greater proportion of clay minerals. Fine-grained terrigenous sedimentary rocks having abundant proportions of quartz, and commonly also of feldspar particles of silt size (Table 2-1) and that by virtue of the abundance of these particles closely resemble sandstones, are commonly referred to as siltstones. Siltstones generally lack fissility.

Fissility is a characteristic of the depositional fabric that is maintained only when burrowing infauna is absent. A fissile shale is one in which burrowing organisms were absent in the depositional environment. According to the definition used by some geologists, a shale must display fissility; more generally, the term shale defines all very fine-grained sedimentary rocks (rocks with particles in the clay- to silt-size range) that contain substantial amounts of clay minerals, whether these rocks are fissile or not. On surface exposures, where weathering has taken place, it is easy to distinguish fissile from non-fissile fine-grained rocks. In subsurface cores or samples, fissility is less well displayed. The terms mudstones, claystones, and argillites have been employed for shales lacking fissility; argillites are more indurated shales than mudstones and claystones. Recall that in the Dunham classification, discussed in Exercise 7, the term mudstone refers to a specific kind of limestone.

References Dealing with Shales

Alling, H. L., 1945, Use of microlithologies as illustrated by some New York sedimentary rocks: Geol. Soc. America, Bull., v. 56, p. 737-755.

Byers, C. W., 1974, Shale fissility: Relation to bioturbation: Sedimentology, v. 21, p. 479-484.

Dunbar, C. O.; and Rodgers, John, 1957, Principles of stratigraphy: New York, John Wiley & Sons, 356 p.

Pettijohn, F. J., 1975, Sedimentary rocks, 3rd Ed.: New York, Harper & Row, Publishers, 628 p.

Picard, M. D., 1971, Classification of fine-grained sedimentary rocks: Jour. Sedimentary Petrology, v. 41, p. 179-195.

Potter, P. E.; Maynard, J. B.; and Pryor, W. A., 1980, Sedimentology of shale - study guide and reference source: New York, Springer-Verlag, 306 p.

Sandstones

Sandstones are terrigenous sedimentary rocks in which sand-size particles dominate (Table 2-1). As with limestones, sandstones can be classified on the basis of (1) particles, (2) matrix, and (3) cement. Particles are mostly those derived from the lands, hence are terrigenous, but pyroclastic and even calcium-carbonate particles may likewise be present in a sandstone. In Exercise 1, we discussed in some detail the kinds of particles that compose sands and sandstones. Basically, the essential particles of sandstones are (1) quartz, (2) feldspar, and (3) rock fragments. As we have seen in Exercise 1, many other kinds of particles occur in sands and sandstones, but those others are not considered essential in naming and classifying sandstones. Quartz occurs in several varieties, and feldspars and rock fragments are of various kinds. The matrix is the mud: fine-grained physically deposited material that occurs mostly as clay minerals and as quartz. Cement is the chemically precipitated filling of originally void spaces.

We distinguish various kinds of sandstone, depending on particles, matrix, and cement. Sandstones devoid of matrix have cement between the particles. Such cement is equivalent to the spar of limestones. As with the micrite of limestones, the argillaceous matrix of sandstones is considered to reflect (1) availability of mud and (2) a low-energy regime in the depositional environment. However, also as with the multiple origins of the micrite in limestones, a fine-grained matrix in a sandstone may filter postdepositionally into the pores among sand-size particles or clay minerals may grow in pore spaces during diagenesis. The end result looks the same; a clay-mineral matrix cannot be distinguished from diagenetically formed clay minerals which are, in fact, cement. Unfortunately, we do not yet know the relative abundance in sandstones of each of (1) mechanically deposited clay-mineral matrix, (2) fine-grained matrix filtered postdepositionally into pores, and (3) clay-mineral cement.

Depending on the presence or absence of argillaceous matter, we recognize two end-member groups of sandstones: (1) argillaceous sandstones (sandstones containing about 15 percent or more of clay-size material); and (2) ordinary sandstones, which contain less than about 15 percent clay-size material. The value of 15 percent is an arbitrary boundary; in a qualitative study of a sample, the obvious presence of a fair amount of clay classifies it as argillaceous. Each of these two groups, (1) argillaceous sandstones, and (2) ordinary sandstones, is divided into three kinds depending on the proportions of quartz, feldspar, and rock fragments. Thus, as indicated in Table 8-1, we recognize as end members six kinds of sandstones: (1) quartz sandstones, (2) feldspar sandstones, (3) rock-fragment sandstones, (4) argillaceous quartz sandstones, (5) argillaceous feldspar sandstones, and (6) argillaceous rock-fragment sandstones.

End-member sand-size particles	Idealized end-member rock names	
	< 15% matrix	> 15% matrix
Quartz	Quartz sandstone	Argillaceous quartz sandstone
Feldspar	Feldspar sandstone	Argillaceous feldspar sandstone
Rock Fragments	Rock-fragment sandstone	Argillaceous rock-fragment sandstone

Classifying and Naming Sandstones
(Friedman and Sanders, 1978, Table 7-4, p. 189)

Table 8-1

These end-member names would be used only in those examples when all the sand-size material consists exclusively of quartz, feldspar, or rock fragments, respectively. Although sandstones consisting exclusively of quartz are not unusual, few examples are known in which the sand-size debris consists entirely of feldspar or of rock fragments. The sand-size debris of most sandstones includes chiefly quartz, but with the quartz may be variable amounts of feldspar or feldspar and rock fragments. These mixtures are named by incorporating the names of all the kinds of sand-size debris present, listed in increasing order of abundance. Thus, if a sandstone contains 70 percent quartz, 20 percent feldspar, and 10 percent rock fragments, we would name it a rock fragment-feldspar-quartz sandstone. The presence of the two modifiers preceding the word quartz indicates that the rock is not a pure quartz sandstone. If sufficient detrital matrix is present for the adjective argillaceous to be used, then "argillaceous" always precedes the names based on kinds of sand-size particles. Thus, a rock containing 20 percent of fine-grained detrital matrix and the proportions of sand-size debris as above would be named an argillaceous rock fragment-feldspar-quartz sandstone.

Although classifying and naming sandstones are descriptive procedures, just as with limestones, genetic overtones have been implied. It has been generally considered that the presence or absence of clay-size particles relates to the level of energy in the original depositional environment. However, as we have just pointed out, other factors, such as the availability of mud in the depositional environment, postdepositional filtering of clay minerals into the sandstone, and diagenetic formation of clay-mineral cements within the sandstone, may almost void the interpretation that clay-size particles reflect energy alone. The kinds of feldspars or rock fragments reflect (1) the source terrain or provenance from which the particles in a sample of sandstone have been derived, and (2) their preservation during diagenesis. As we have seen in Exercises 1 and 5, much useful geologic information can be extracted from a study of the kinds of particles that compose a sandstone.

Earlier the dogma was widespread that certain kinds of sand-size particles always and invariably became associated with a particular amount of matrix, which we now realize may not be matrix at all. The originator of this idea was Paul D. Krynine, the pioneer sedimentary petrographer who provided the modern impetus in naming and classifying sandstones. Krynine's original article is a classic in geology; it combines astute observation in the field and of samples with a deep insight and appreciation for the fundamental principles of geology. Krynine's contribution is felt to this day, yet much misunderstanding resulted from the application of his ideas. In his endeavor to train geologists in observing, describing, and interpreting sedimentary rocks, he leaned on the "simple and logical nomenclature of igneous petrography." In his classification of sandstones, Krynine first set out the "petrographic classification of igneous rocks" and then proceeded "just as in igneous rocks" with the "petrographic classification of sediments." He introduced as end members of his classification of sandstones, in a triangular diagram, the terms quartzite, graywacke, and arkose; the use of the last two of these terms has since led to misunderstanding and confusion. The earliest description of graywacke dates from 1789, and the term arkose was introduced in 1823. Both terms had had a checkered history even before Krynine included them in his classification. According to Krynine, a graywacke is a sandstone composed of angular quartz (and chert) particles and abundant metamorphic rock fragments, with little or no cement and feldspar, and containing more than 12 to 17 percent micas and chlorite (either in the clay matrix or as metamorphic rock fragments). Krynine felt that the mica and chlorite between the particles were indeed a mechanically deposited matrix and represented poor sorting; we now know that this basic assumption can no longer be assumed to be correct. To avoid the term graywacke, some authors substituted the term wacke. In his classification, Krynine used the term arkose as a sandstone with more than 30 percent feldspar.

In our scheme of naming sandstones, a rock defined by Krynine as a graywacke would be named an argillaceous rock fragment-quartz sandstone if it contained no feldspar; and an argillaceous rock fragment-feldspar-quartz sandstone, if it contained feldspar in greater abundance than rock fragments; or an argillaceous feldspar-rock fragment-quartz sandstone, if the rock fragments were more abundant than the feldspar. The arkose of Krynine would be named by us as a feldspar-quartz sandstone, if the quartz is more abundant than the feldspar, but a quartz-feldspar sandstone if feldspar abundance exceeds quartz abundance.

In his classic paper, Krynine's stated purpose was to introduce a purely descriptive classification. Krynine believed that the name of a sandstone should reflect its mineral composition (he probably meant the various kinds of particles that compose the rock) and its texture. However, as we have seen with limestones, a purely descriptive classification may have genetic implications. Krynine recognized this, but this is where he failed. He claimed that "it so happens, that this classification is also genetic, since every major sedimentary series is related to a definite degree of intensity of deformation of the earth's crust (diastrophism), operating during the deposition of this rock series." Krynine's emphasis on diastrophism made his classification less viable. You will recall that the particles reveal provenance and the clay fraction (1) provenance, (2) availability of mud in the depositional environment, (3) the level of energy of the depositing medium, and (4) diagenetic formation of clay-mineral cement; the effect of diastrophism is only indirect. To Krynine and to some of his followers, genetic implications have become attached to the terms graywacke and arkose. Therefore, we do not adopt these two terms in our approach to the classification of sandstones. Although we have given Krynine's definitions of these two terms, we recommend that they be considered as historical terms only and be dropped from further usage in describing and classifying sandstones.

Since Krynine proposed his original classification of sandstones, approximately 50 new classifications, based mostly on Krynine, have been proposed in more than 10 countries and in seven languages. Most of these classifications are variants on his theme. Since we favor simple words in naming and classifying sandstones and use visual estimates for compositions, we have adopted in this chapter the simple scheme of Table 8-1 rather than the elaborate classifications based on Krynine or variants thereof. This classification of sandstones parallels those for limestones, which are outlined in Exercise 7. Some geologists make quantitative analyses by point counting the distribution of the various kinds of particles in thin sections under the petrographic microscope. Those who do so compute their three kinds of particles on the basis of 100 percent and plot them on a triangular diagram (Fig. 8-1). In their work, these operators have more detailed information than can be shown on a triangular diagram, such as the amounts of the various kinds of feldspars and rock fragments. The purpose of the triangular diagram is to compare and contrast the composition of sandstones from various strata in one local area, in one basin, or even on a worldwide scale.

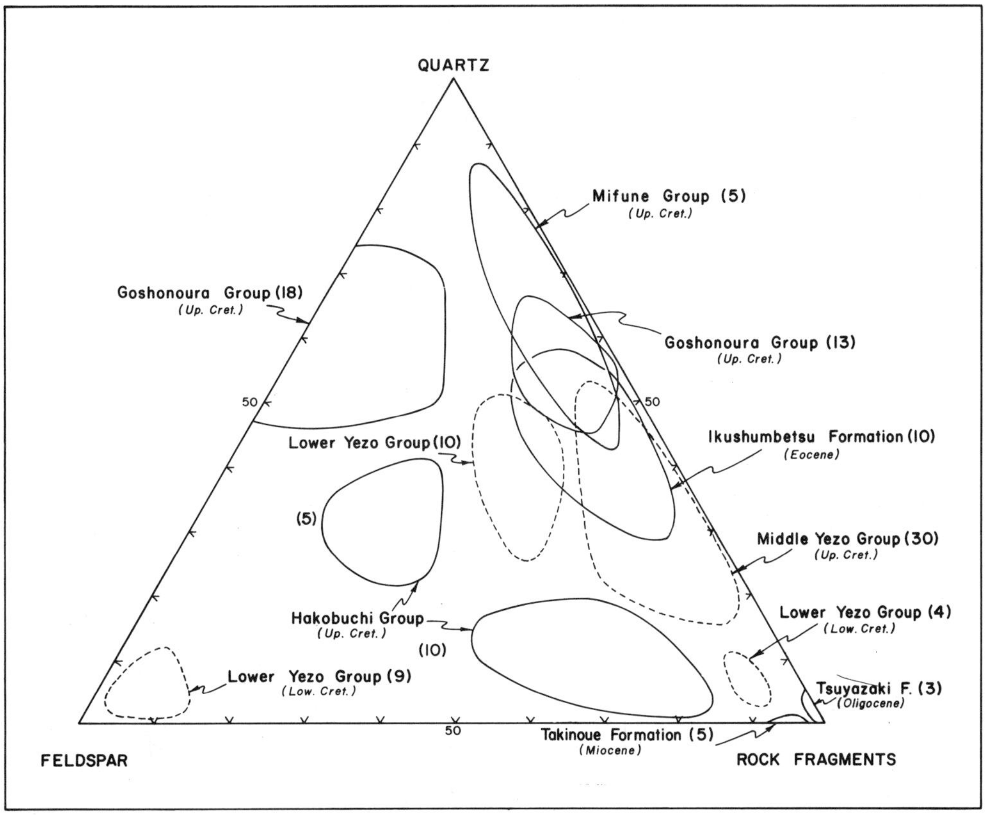

Triangular diagram showing the composition of various kinds of sandstones from selected strata in Japan. Lines enclose fields of scatter of many specimens (number shown in parentheses) point counted for each group of rocks. (H. Okada, in Friedman and Sanders, 1978, Fig. 7-19, p. 190.)

Figure 8-1

References Dealing with Sandstones

Boswell, P. G. H., 1960, The term graywacke: Jour. Sedimentary Petrology, v. 30, p. 154-157.

Dott, R. H., Jr., 1964, Wacke, graywacke, and matrix - What approach to immature sandstone classification?: Jour. Sedimentary Petrology, v. 34, p. 625-632.

Folk, R. L., 1956, The role of texture and composition in sandstone classification: Jour. Sedimentary Petrology, v. 26, p. 166-171.

Klein, G. deV., 1963, Analysis and review of sandstone classifications in the North American geological literature, 1940-1960: Geol. Soc. America, Bull., v. 74, p. 555-576.

Krynine, P. D., 1948, The megascopic study and field classification of sedimentary rocks: Jour. Geology, v. 56, p. 130-165.

Pettijohn, F. J., 1960, The term graywacke: Jour. Sedimentary Petrology, v. 30, p. 627.

Pettijohn, F. J.; Potter, P. E.; and Siever, R., 1972, Sand and sandstone: New York, Springer-Verlag, 618 p.

Conglomerates and Sedimentary Breccias

The coarse-grained terrigenous rocks are formed by the lithification of gravel. Two varieties are recognized based on the degree of rounding of the particles. Conglomerate is defined as a coarse-grained terrigenous sedimentary rock formed by the lithification of rounded gravel. Sedimentary breccia is defined as a coarse-grained terrigenous sedimentary rock formed by the lithification of angular gravel. In conglomerates and sedimentary breccias, more than 30 percent of the large particles exceed ($-1\ \phi$) in diameter. The particles may consist of pebbles, cobbles, boulders or mixtures of these sizes (Table 2-1).

Conglomerates and sedimentary breccias may be named and classified by (1) the proportion of gravel-sized particles, (2) the kind of matrix, and (3) the kinds of gravel-sized particles. Conglomerates have also been assigned names by origin, for example, fanglomerate and tillite.

According to proportion of gravel-sized particles and the kinds of matrix, a sample containing (1) >80 percent pebbles, cobbles, and/or boulders is termed a conglomerate proper, and (2) one having between 30 and 80 percent of such coarse particles is (a) a sandy or arenaceous conglomerate, or (b) a shaly or argillaceous conglomerate. The matrix between the coarse particles in a conglomerate may also be calcareous or sideritic. As we have seen in the above definition, a rock is named a conglomerate with only about 30 percent coarse particles; that is, the sandy or shaly material may exceed that of the gravel-sized particles in bulk volume. The reason for lowering the limit for gravel-sized particles below 50 percent and still defining the rock as conglomerate is the significance attached to the presence of such coarse particles. The proportion of gravel is a function of (1) the highest current speed at the time of deposition, and (2) the availability of particles of such coarse size. Moreover, empirical experience has shown that in outcrop or in hand samples geologists will name a rock a conglomerate with the proportion as low as, and even lower than, 30 percent gravel-sized particles. Pebbles just stand out like plums in a pudding, and accordingly the rock is named conglomerate.

Based on the variety of pebbles, cobbles, and/or boulders composing them, we can classify conglomerates into two kinds: (1) conglomerates consisting only of pebbles, cobbles, and/or boulders of a single kind of rock (such as one or various varieties of chert and quartzite or of other rock) or oligomictic conglomerates; and (2) conglomerates containing pebbles, cobbles, and/or boulders of many kinds of rock, known as polymictic conglomerates. The term mictic derives from the Greek word miktos, meaning mixed; oligo- comes from oligos, meaning few, and poly- from polys meaning many, both terms also Greek.

An example of an oligomictic conglomerate is the conglomerate of the Cardium Formation (Middle Upper Cretaceous) of Alberta, Canada (Fig. 8-2). As with many other oligomictic conglomerates, the Cardium Conglomerate contains pebbles that are resistant to wear and decomposition: quartzite, chert, siliceous shale, and radiolarian chert; some quartzite pebbles have been brecciated and healed with vein quartz. The pebbles are markedly worn and well rounded. Their surfaces are smooth and polished, and many exhibit indentations or pits. The color of individual chert pebbles includes white, gray, green, blue, and black. Pebbles generally make up an estimated 65 to 85 percent of the bulk composition of this conglomerate. Ellipsoidal pebbles are locally oriented with their axes of elongation parallel to the bedding. The finer-grained particles between the pebbles consist of quartz and chert of sand and silt size as well as of small pebbles of a size that falls close to the sand-size boundary.

Some geologists claim that oligomictic conglomerates are derived from a deeply weathered terrain in which by long transport all but the most resistant pebbles have been eliminated. Among the kinds of rock most prone to wear and tear and early elimination are limestone clasts, which do not survive extensive abrasion. Yet numerous examples of oligomictic conglomerates consisting of limestone clasts exist in the rock record. Most of

Core from borehole showing well-rounded chert pebbles in oligomictic conglomerate. Cardium Formation (Cretaceous), Pembina, Alberta, Canada. (Friedman and Sanders, 1978, Fig. 7-20, p. 192.)

Figure 8-2

these conglomerates accumulate on the downthrown side of an intermittently active fault where the upthrown side contains limestones. To offset the term tillite, which has genetic connotations, the term diamictite has been proposed for a nonsorted, noncalcareous, terrigenous sedimentary rock composed of sand-size and/or larger particles dispersed through a fine-grained matrix. Many, if not most, diamictites are conglomerates. Diamictites originate by landslides, earth flows, mud flows, solifluction, ice rafting, subaqueous slumping and sliding, and/or by glaciers. Some diamictites are tillites, others are not. Before a diamictite became lithified, it was a diamicton.

Exposure of diamictite (tillite) showing pebbles and boulders, mostly gneisses, embedded in sandy and argillaceous matrix. Large boulder in center of photograph is an erratic of gneiss. Adelaide System (Precambrian), Sturt Gorge near Adelaide, South Australia. (Friedman and Sanders, 1978, Fig. 7-21, p. 193.)

Figure 8-3

References Dealing with Conglomerates and Sedimentary Breccias

Flint, R. F.; Sanders, J. E.; and Rodgers, J., 1960, Diamictite, a substitute term for symmictite: Geol. Soc. America, Bull., v. 71, p. 1809-1810.

Hubert, J. F., 1960, Petrology of the Fountain and Lyons formations, Front Range, Colorado: Colorado School Mines, Quart., v. 55, 242 p.

Pettijohn, F. J., 1975, Sedimentary rocks, 3rd Ed.: New York, Harper & Row, 628 p.

Twenhofel, W. H., 1947, The environmental significance of conglomerates: Jour. Sedimentary Petrology, v. 17, p. 119-128.

Pyroclastic Rocks

Sedimentary petrographers have neglected the study of pyroclastic rocks in part because the particles are of igneous origin and in part because most early classical studies in sedimentology took place in areas removed from active volcanism. Furthermore, because of their chemical instability, pyroclastic rocks are subject to rapid diagenetic alteration, which makes those of Paleozoic and Precambrian and even those of Mesozoic age difficult to identify. As part of the science of paleovolcanology, igneous petrographers gave these rocks brief attention. This attention is evident by the names, which truly reflect the "specious" species concept; examples are piperno, piperini, eutaxite, pseudoignimbrite, Kai-isi, sillar, wilsonite, owharoite, ignitorrent, ignispumite, and ignimyssite, to name just a few in the chaotic literature of pyroclastic rocks. We shall not discuss these names; our purpose in giving them here is to illustrate the danger inherent in naming rocks by the species concept.

Volcanoes and their products, both on land and in the sea, are the major source of particles for pyroclastic rocks, especially in areas of tectonic activity and in island arcs. Particles originating by explosive ejection from volcanic vents spread out through great areas. On the continents, pyroclastic rocks far exceed the volume of lava.

Pyroclastic rocks may be classified by (1) size, (2) the kinds of particles composing them, and (3) origin. We shall use size and composition to classify them, but shall avoid origin. On the basis of size, pyroclastic particles are organized into three categories: (1) particles larger than 64 mm; (2) particles between 64 and 2 mm; and (3) particles smaller than 2 mm.

Particles larger than 64 mm are further divided into two groups based on shape. These are bombs, particles, 64 mm in diameter, ejected as liquid lava that solidified during flight and hence commonly have twisted forms and concentric structure, and blocks, particles, 64 mm in diameter, ejected from a volcano as solids and consisting of angular fragments of volcanic rock.

Particles in the intermediate size range are called lapilli, which are defined as pyroclastic particles 2 to 64 mm in diameter. Particles in the fine size range are named ash and defined as pyroclastic particles 2 mm in diameter.

These three size groups form the basis for classifying and naming pyroclastic rocks as follows. Pyroclastic rocks composed mostly of bombs are agglomerates. Pyroclastic rocks composed mostly of blocks are volcanic breccias. Pyroclastic rocks composed mostly of lapilli are lapilli tuffs. And pyroclastic rocks composed of ash are tuffs in the strict sense of the word.

Classifying a pyroclastic rock on the basis of composition, that is, by the kinds of particles composing it, is usually difficult. Many rocks are very fine grained or glassy, and others are glassy in part. Where glass is absent and the particles are coarser grained, they may be recognized in a thin slice under the petrographic microscope. The particles that are not glass are volcanic rock fragments or crystals of various minerals, as shown in Figure 1-2. According to composition, pyroclastic rocks have three end members, (1) rock fragments, (2) crystals, and (3) glass. In a thin slice under the petrographic microscope, an operator estimates (or point counts) the relative proportions of these three end members. If by size a rock is a tuff and rock fragments exceed glass, this tuff may be described as a glass, rock-fragment tuff. End members are listed in order of increasing abundance; there may be a crystal, rock-fragment, glass tuff. Adjectival modifiers may be added.

Because of the chemical reactivity and instability of their particles, pyroclastic rocks are highly susceptible to diagenetic alteration. Glass alters to clay minerals, especially to montmorillonite, zeolites, chalcedony, opal, quartz, or to a microcrystalline particle, which in a thin slice under the petrographic microscope resembles chert. Glass is commonly found only in mid-Tertiary and younger rocks. Even zeolites are absent from lower Paleozoic rocks; they probably alter to feldspars. An alteration product of tephra is bentonite, a plastic clay or shale composed for the most part of the clay mineral montmorillonite that formed by the alteration of tephra.

One kind of tuff that carries a special name is ignimbrite or welded tuff, a nonsorted pyroclastic rock deposited from a nuée ardente while the particles were still in a plastic condition. Ignimbrites may cover large areas and attain considerable thicknesses.

References Dealing with Pyroclastic Rocks

Kittleman, L. R., Jr., 1973, Mineralogy, correlation, and grain-size distributions of Mazama tephra and other postglacial pyroclastic layers. Pacific northwest: Geol. Soc. America, Bull., v. 84, p. 2957-2980.

__________, 1979, Tephra: Scientific American, Dec., pp. 160-177.

Koch, A. J.; and McLean, Hugh, 1975, Pleistocene tephra and ashflow deposits in the volcanic highlands of Guatemala: Geol. Soc. America, Bull., v. 86, p. 529-541.

Lajoie, Jean, 1979, Facies models 15. Volcaniclastic rocks: Geoscience Canada, v. 6, p. 129-139.

McBirney, A. R.; and Murase, T., 1970, Factors governing the formation of pyroclastic rocks: Bull. Volcanol., v. 34, p. 372-384.

Walker, G. P. L., 1971, Grain-size characteristics of pyroclastic deposits: Jour. Geology, v. 79, p. 696-714.

Procedure

Using the specimens of extrabasinal and pyroclastic rocks provided for study, carefully determine the compositional and textural character of each one and record your observations on Form 8-1. Examination should include use of hand-lens, dilute HCl, binocular microscope and petrographic microscope, as appropriate.

After recording observational data, name each specimen.

Exercise 8
Name ______________________
Laboratory Section ______________
Date ______________________

REPORT ON CLASSIFICATION OF EXTRABASINAL AND PYROCLASTIC ROCKS

Name	
Texture	
Composition	
Speci-men #	

Form 8-1

Speci-men #	Composition	Texture	Name

Form 8-1 (continued)

Geologists, in their all but closed conversation, inhabit scenes that no one ever saw, scenes of global sweep, gone and gone again, including seas, mountains, rivers, forests, and archipelagoes of aching beauty rising in volcanic violence to settle down quietly and then forever disappear - <u>almost</u> disappear.

John McPhee

EXERCISE 9
FACIES ANALYSIS

Introduction

As you are aware, in physics and chemistry, known procedures are followed in carefully controlled experiments to see what unknown products will result. After the products have formed, they are directly observed and related exactly to the formative processes. The reasoning is from process to product. By contrast, in geology the rocks are known products and the formative processes are the unknowns. Geology is thus unique among the sciences in that the geologist starts with the result (the rocks) and asks, in effect, "What was the experiment?"

Exercises 1 through 8 dealt with the compositional and textural nature of sedimentary particles and the myriad combinations in which such particles may accumulate in nature. These particle assemblages (rocks and sedimentary deposits) are the products of sedimentary processes operative in natural settings called depositional environments. This exercise provides an introduction to facies analysis, the interpretation of the rock record for purposes of reconstruction of the physical, chemical and biological processes that were responsible for the original sediments.

Depositional Environments and Facies

The geologist must be familiar with modern sediments and laboratory experiments with sediments where close relationships between processes and products have been established. It is important to recognize, however, that in dealing with natural processes, geologists are confronted not with the products of processes that operated in isolation but rather with products of associated processes that operated collectively in depositional environments. Deposits which evolve in specific depositional environments, such as intertidal flats, in desert lakes, or in salt marshes, are considered to be sedimentary facies. The term facies refers to the lithologic and biologic characteristics of a sedimentary deposit imparted by the processes operating in the depositional environments. Some examples of facies in the rock record are fluvial facies, eolian facies, and tidal facies, to mention just a few. Because many depositional environments exist simultaneously, and because the geographic extent of any given environment is not infinite, the sediments deposited simultaneously change facies from place to place; that is, various facies interfinger with one another. Accordingly, a beach facies may interfinger landwards with a coastal-dune facies, and this, in turn, may pass into fluvial and other nonmarine facies. In the seaward direction, a beach facies may interfinger with various shallow marine facies, which ultimately may interfinger with deep-sea facies. The modern usage of facies with its specific emphasis on facies changes stresses this interfingering.

A given facies is deposited only within the area occupied by its specific depositional environment. In other words, at any one time, geography determines the horizontal dimensions of each facies. This being true, how is it possible for a facies to eventually occupy a wider area than that occupied by its depositional environment? The answer to this is that the boundaries of the environments shift with time. As these boundaries migrate, the characteristic deposits of the environment are even more widely distributed. In this way, many facies are spread laterally with time and are "stacked" vertically on top of one another.

There are three general groups of depositional environments: (1) nonmarine, (2) marine, and (3) transitional between nonmarine and marine. Table 9-1 is a tabulation of those depositional environments which are important for purposes of this exercise in facies analysis.

Group	Environment	Sub-Environment
Nonmarine	Alluvial	Channel Overbank Marsh and Swamp
Transitional	Tidal	Channel Flat
Marine	Nearshore Offshore	

Sedimentary Environments Considered in Exercise 9

Table 9-1

There are, of course, many other environments which in some geographic areas and some parts of the rock record are very important, but which do not happen to be represented in this exercise. Detailed information on the characteristics of all the recognized depositional environments and facies may be found in the following references:

Braunstein, Jules, 1974, Facies and the reconstruction of environments: Tulsa, American Assoc. Petroleum Geologists Reprint Series No. 10, 223 p.

Curtis, D. M., Compiler, 1978, Depositional environments and paleoecology: Environmental models in ancient sediments: Tulsa, Society of Economic Paleontologists and Mineralogists Reprint Series No. 6, 240 p.

Dunbar, C. O.; and Rodgers, John, 1957, Principles of stratigraphy: New York, John Wiley & Sons, 356 p.

Friedman, Gerald M.; and Sanders, John E., 1978, Principles of sedimentology: New York, John Wiley & Sons, 553 p. Part V (Modern sedimentary environments and their products in the rock record, p. 195-401).

Matthews, R. K., 1974, Dynamic stratigraphy: Englewood Cliffs, New Jersey, Prentice-Hall, 370 p. Part III (Sedimentary environments from the mountains to the deep sea, p. 135-318).

Reading, H. G., ed., 1978, Sedimentary environments and facies: New York, Elsevier, 557 p.

Selley, R. C., 1970, Ancient sedimentary environments: London, Chapman and Hall, Ltd., 237 p.

A depositional environment is considered as being defined by a set of values of physical, chemical, and biological variables that correspond to a particular geomorphic unit, such as an alluvial flood plain, a lagoon, or a tidal flat. The interplay of processes operating within a given environment gives the sedimentary deposit distinctive physical, chemical, and biological characteristics. The characteristics are variable in the extent to which they are preserved in the rock record. Physical characteristics (such as sedimentary structures) are commonly well preserved, but chemical characteristics may be altered by diagenetic processes, and biologic elements may in large part be lost by selective destruction due to oxidation, reduction, or chemical solution.

In order to recognize sedimentary facies in the geologic record it is necessary to use multiple recognition criteria including: (1) sedimentary structures, (2) lithology, (3) geometric relationships of rock units, and (4) character of biologic content.

The following tables include information on the specific characteristics of the rock facies which evolved in the depositional environments outlined in Table 9-1. These characteristics are typical for sedimentary rocks deposited as a clastic wedge at the margin of a basin where the source area was a nearby orogenic zone dominated by low-grade metamorphic rocks.

	Channel	Overbank
Lithology	Gray and greenish gray, fine to medium-grained, immature argillaceous rock fragment-quartz sandstone containing abundant carbonaceous debris and green shale pebbles (shale pebble conglomerate), especially in lower 0.5 to 1 m of sandstone body.	Red and green, locally mottled siltstones, mudstones and shales with sporadic dark gray, locally bituminous lenses and very sparse thin, greenish tan, highly calcareous siltstone beds.
Texture	Variable; poorly sorted with abundant "fines," commonly becoming finer-grained upward in sandstone body.	Consistently of silt and clay grain-size with very small admixtures of very fine sand grains.
Sedimentary Structures	Tabular and trough cross-bedding ubiquitous, commonly with decreasing foreset thickness and inclination upward in sandstone body. Parting lineation common. Current ripple marks locally well developed.	Very thin parallel laminae which generally are obscure on even slightly weathered outcrops; locally bedding is shaly. Local occurrence of mud-cracks.
Geometry	Basal contact truncates strata of underlying unit; upper contact gradational into siltstone of overbank facies. In cross-section sandstone body may interfinger with overbank facies or may be lenticular. Many bodies are laterally extensive. Where abundant subsurface control is available, bodies would be elongate and exhibit bifurcating patterns.	Facies occurs interbedded with channel sandstones. Basal contact with sandstone body gradational; upper contact a sharp, channelled disconformity. Interfingers laterally with channel facies.
Associations	Sandstone bodies of facies occur as cyclic, "multistory" interbeds in red and green and locally dark gray and black, highly organic siltstones and shales of overbank and marsh facies.	See Geometry. Green coloration occurs in red siltstones and mudstones as mottles, lenses and very thin, persistent layers.
Miscellaneous	Outcrop surfaces commonly iron-stained. Iron sulfates (melanterite and jarosite) commonly occur on organic, pyrite-rich lenses. Only contained fossils are plant material, sparse freshwater bivalves and fish plates.	Only fossils are plant material. Locally burrows, green burrow mottling and yellow-green, very slightly calcareous plant stigmaria are abundant.

Summary of diagnostic characteristics of the alluvial facies.

Table 9-2

	Flat	Channel
Lithology	Medium dark gray, finely micaceous, muddy silt-stone and very subordinate medium gray, very fine-grained sandstone.	Medium gray, fine-grained, immature rock fragment-quartz sandstone containing abundant plant materials; coquinoid sandstone and coquinite with spiriferid brachiopods; polymictic conglomerate with pebbles of gray, light green, and olive silt-stone, greenish gray quartzite and red siltstone and shale.
Texture	Very fine to fine grade-sizes.	Grain-size decreases upward in rock unit from conglomerate lenses in base to very fine-grained sandstone at top. Sorting is generally poor.
Sedimentary Structures	Featureless to very fine cross-laminated (flaser and ripple bedding). Local, well developed polygonal mud-cracks and oscillation ripple marks.	Tabular and trough cross-bedding ubiquitous in basal part of sand-stone, grading upward into generally horizontal bedding at top. Parting lineation common in upper part of body.
Geometry	Basal contact gradational from sandstones of inferred tidal or near-shore bar origin. Upper contact disconformable with overlying tidal channel sandstone or nearshore bar sandstone.	Basal contact truncates strata of underlying inferred tidal flat facies. Upper contact gradational into fine-grained rocks of tidal flat or other facies. In cross-section may be lenticular or interfinger with tidal flat facies. In modern environments long axis of sand body generally perpendicular to shore-line.
Associations	Interbedded with sand-stones of inferred near-shore bar and tidal channel origin.	Sandstone bodies occur as channel deposits in tidal flat facies. Basal lag concentrate is commonly coquinite or coquinoid sandstone.
Miscellaneous	Contains abundant plant material and common allochthonous brachiopods.	Internal sedimentary structural configuration the same as that of alluvial channel facies. Sand-stones contain local large burrow structures and allochthonous marine fossils

Summary of diagnostic characteristics of the tidal facies.

Table 9-3

Lithology	Predominantly medium gray and olive gray sandstone. Although of low maturity index, relatively "clean" compared to tidal and alluvial sandstones. Variability low. Subordinate medium gray siltstone containing abundant wood material.
Texture	Very fine-grained sand. Maturity index, although low, is higher than in tidal and alluvial sandstones. Variability of texture low.
Sedimentary Structures	Very shallow cross-bedding in sandstones. Oscillation and current ripple marks locally abundant. Flow rolls locally well-developed, especially at distal margin of nearshore facies. Wave and current directional indicators suggest variable direction of sediment transport.
Geometry	Bar sandstones lenticular in cross-section. Basal contacts on offshore siltstones are of both transitional and abrupt types.
Associations	Nearshore sandstones interbedded with siltstones and shales of lagoon facies. General gradational fining-upward beneath transgressive shale tongue and coarsening-upward above tongue. Sandstones of nearshore facies grade shoreward into sandstones of tidal facies and basinward into siltstones of offshore facies.
Miscellaneous	Abundant brachiopods and pelecypods, many composed of original calcite. Some brachiopods occur in coquinoid, conglomeratic lenses. Vertical burrow structures locally abundant. Finely mascerated plant material ubiquitous.

Summary of diagnostic characteristics of the nearshore facies.

Table 9-4

Lithology	Medium gray, slightly calcareous to calcareous siltstone grading basinward into argillaceous micrite and biomicrite. Distal offshore strata are brownish black and very dark gray due to high content of very fine carbonaceous and bituminous fragments and disseminated iron sulfides.
Texture	Very fine (silt) grade sizes. Variability of texture low.
Sedimentary Structures	Bedding poor and very thin to laminated. Micro-cross-lamination ubiquitous. Wave and current directional indicators suggest greater constancy of sediment transport direction than in near-shore facies. Distal offshore rocks are massive and horizontally bedded with very fine parallel laminae. Weathered outcrops are very fissile.
Geometry	Distal offshore facies is a tongue penetrating shoreward into more proximal offshore rocks.
Associations	In landward direction, offshore siltstones interfinger with and grade into nearshore bar sandstones; in basinward direction grade into argillaceous micrite. Distal offshore strata grade downward into micrite and upward into more proximal offshore siltstones.
Miscellaneous	Abundant pelecypods and brachiopods, some composed of original calcite. Very fine plant material ubiquitous. Proximal offshore clastics contain horizontal burrow structures. Micrites contain well-developed "mottled" burrow structures. Distal offshore shale contains sparse fossils.

Summary of diagnostic characteristics of the offshore facies.

Table 9-5

Materials

The following materials will be needed for this exercise:

Metric scale
One sheet of 30 x 60 cm white drawing paper
Drawing pencils (2H, 4H)
Colored pencils

Procedures

The objective of this exercise is to introduce you to facies analysis. You are provided with field data in the form of a tabular description of the rock strata present in seven stratigraphic sections (Tables 9-6 through 9-12). You will plot this information on a stratigraphic cross-section and then, referring to Tables 9-2 through 9-5 as a guide, determine the spatial relationships of the sedimentary facies. As indicated earlier, these are rocks that evolved as a clastic wedge at the margin of a basin where the source area was a nearby orogenic zone dominated by low-grade metamorphic rocks. Clastic influx into the basin was high, so that the sorting of the rocks is generally quite poor.

The abbreviations used for the stratigraphic section descriptions in Tables 9-6 through 9-12 are:

gy. - gray, gn. - green, rd. - red, lt. - light,
gn.-gy. - greenish gray
f. - fine, m. - medium, c. - coarse, g. - grained
vy. - very, bdd. - bedded, lam. - laminated, calc. - calcareous,
slty. - silty, shly. - shaly, sndy. - sandy
Contacts: ———————— sharp, —— —— —— gradational
SS - sandstone, SH - shale, SLTST - siltstone,
MDST - mudstone

Top of Section DD 78-325

1 m	gy., v.f.g. SS
25 m	gy. SH, in upper part sandy and calc., grading downward into slty. SH
11 m	gy., shly. BIOMICRITE, burrowed, in part w horn corals and w occ. oscillation ripple marks
0.3 m	gy., INTRASPARITE w common SH chips, abundant oöids and sparse crinoid fragments
0.6 m	gy., calc., shly. SLTST, w interlam. of v.f.g. BIOSPARITE

Base of Section

NOTE: Use a point 8 m above the base of this section as datum for your stratigraphic cross-section.

Section 1

Table 9-6

25
11
1
37 38

11.9
8
3.9M

Top of Section DD 79-1

12 m	gy. SH w burrow tubes and v.f. plant frags. grading downward through gy., vy. shallowly cross-lam. SLTST to gy., vy. shallowly cross-lam., slightly calc., v.f.g. SS
11 m	gy., cross-lam. and shly. SLTST w ubiquitous v.f. plant frags. and w vy. f. burrows and pelecypod casts.
13 m	gy., vy. shallowly cross-lam. SLTST and v.f.g. SS, in part burrowed, in part calc., w occ. oscillation ripple marks and brush casts
15 m	gy. vy. shallowly cross-lam. SLTST and v.f.g. SS as above, with occasional pyrite nodules
9 m	gy., vy. shallowly cross-lam. SLTST grading downward into calc. SLTST
17 m	gy., calc. SH and shly. SLTST with common burrow tubes, small horn corals and sparse vy. fine plant fragments
2 m	lt. gy., vy. fossiliferous, in part oölitic, in part highly burrowed, CALCILUTITE

(Datum for stratigraphic cross-section)

gy., shallowly cross-lam., silty v.f.g. SS

Base of Section

Section 2

Table 9-7

Top of Section DD79-2

Thickness	Description	Note
26 m	gy., shallowly cross-lam. f.g. SS grading downward into cross-lam. sndy. SLTST and SLTST, all w sparse v.f. plant fragments, in part w sparse to abundant brachiopod casts, w occ. small calc. or pyrite nodules	TF
11 m	gy., cross-lam., shly. SLTST, w occ. small lenses of calc., v.f.g. SS.	TF
13 m	gy., vy. shallowly cross-lam., f.g. SS w abundant v.f. plant fragments and sparse to very abundant brachiopod casts.	NS
2 m	gy., dense, v.f.g. SS w ball & pillow structure and abundant brachiopod casts.	NS
34 m	gy., shallowly cross-lam., f.g. SS, w sparse to abundant f.g. plant mat'l. & abundant, sometimes large, brachiopod casts; f.g. SS grades downward through gy. v.f.g. SS to darker gy., shly. SLTST, all w abundant v.f.g. plant mat'l., in part w brach. casts and w abundant current & oscill. ripple marks.	NS / OS
15 m	gy., vy. thin bdd. to shaly, in part calc., SLTST w pelecypod casts, v.f. plant fragments and occ. interbeds of vy. thin, vy. calc., v.f.g. SS.	TF
1 m	gy., thin bdd., v.f.g. SS	
2 m	interbedded gy., in part calc., SH and vy. thin bdd. BIOCALCARENITE w horn corals, abundant crinoid fragments and vy.f. plant fragments (crd SS)	NS
	Datum for stratigraphic cross-section	
44 m	gy. SH and vy. thin bdd. SLTST, in part sandy, w vy. thin interbeds of gy., calc., f.g. SS, all w common brachiopod casts and in part w v.f. plant frags., burrows, gastropod casts and crinoid fragments.	NS / TF

Base of Section

Section 3

Table 9-8

Top of Section ————————

4 m	gn.-gy., med. cross-bdd. (foreset thickness > 10 cm) m.-c.g. SS w abundant, in part coarse, plant fragments.
11 m	gy., shallow cross-lam. & thin to vy. thin bdd. SLTST w abundant f.-c. plant frags. & sparse to occ. abundant brachiopod casts.
9 m	gy., thin & vy. thin poorly bdd., v.f.g. SS w occ. SLT lenses, abundant f. plant frags., common to abundant brachiopod casts & occ. burrows.
28 m	olive gy. & lt. olive gy., thin poorly bdd. & finely cross-lam. SLTST w abundant fine plant frags. & common brachiopod casts.
9 m	gy., shallowly cross-bdd., calc., f.g. SS w abundant plant frags. & brachiopod and pelecypod casts, grading downward into gy., thinly bdd. & cross-lam. SLTST.
47 m	gy., thin & shallowly cross-bdd., f.g. SS w abundant f. plant frags. & occ. brachiopod casts.
15 m	gn.-gy., lam., cross-lam. & vy. thin bdd. SLTST w occ. brachiopod casts & abundant f. plant frags.
27 m	gy., thin, well-bdd., f.g. SS w abundant ball & pillow structures in lower 40' of interval, & w abundant f. plant frags., fucoids & brachiopod casts throughout entire interval.
7 m	gy., thin, vy. shallow cross-bdd. SLTST w abundant f. plant frags. & brachiopod casts & occ. fucoids.

Base of Section ————————

NOTE: Use a point 61 m above the base of this section as datum for your stratigraphic cross-section.

Section 4

Table 9-9

Top of Section ——————

1 m	gn.-gy., thin, vy. shallowly cross-bdd., highly calc., m.g. SS w abundant brachiopods composed of $CaCO_3$ & large burrows.
5 m	gn.-gy., thin, vy. shallowly cross-bedded, highly calc., m.g. SS w sltst & sh. pebbles, occ. pyrite nodules & c. plant frags.
6 m	Gy. SLTST w abundant f. plant frags.
8 m	gn.-gy., thin bdd. & cross-lam. to thick cross-bdd., f.g. SS w abundant f. plant frags.
1 m	gn.-gy., thin bdd. & cross-lam., shly., calc., f.g. SS, in part w sh. frags., in part vy. dense & "clean."
24 m	gn.-gy., variably bdd., v.f.g. SS, upper part "clean" & lower part w sh. peb. cglt. lenses, f. plant frags. and occ. abundant fucoids.
12 m	gy. SLTST w v.f.g. SS lenses, w abundant plant frags. & brachiopod casts.
6 m	gn.-gy., vy. thick bdd., f.g. SS w f. plant frags., grading downward into gy., thick bdd. v.f.g. SS w f. plant frags.
1.5 m	gy., thick bdd., f.g. SS interbdd. w gy., fissile SH.
4 m	gn.-gy., med. to thick bdd., f.g. SS w f. to c. plant frags., sparse brachiopod casts & sh. peb. cglt. lenses.

Base of Section ——————

NOTE: Use a point 20 m below the top of this section as datum for your stratigraphic cross-section.

Section 5

Table 9-10

Top of Section

13 m	gy., vy. thin, vy. shallow cross-bdd. & cross-lam. f.g. SS w abundant f. plant frags., grading downward to slty. SS.
3 m	gy. SH & shly. SLTST w occ. pyrite nods.
0.3 m	gy., vy. calc., f.g. SS w abundant large, $CaCO_3$, brachiopods & f. plant frags.
0.3 m	dk. gy. SH w sparse pelecypod casts.
6 m	gn.-gy., thin shallow cross-bdd. & "flaggy," f.g. SS w abundant f. plant frags. & sparse burrows & occ. current ripple marks.
7 m	Interbedded gy., shly. SLTST & vy. calc., cgltc., f.g. SS w abundant $CaCO_3$ brachiopods. Pelecypod casts, f. plant frags. & oscillation & current ripple marks.
2 m	olive gy., vy. thin shallow cross-bdd. f.g. SS w common sh. pebs. & f. to c. plant frags.
3 m	gy., vy. thin, vy. shallow cross-bdd. & cross-lam. f.g. SS w f. plant frags. & occ. ball & pillow structures.
0.3 m	gy., vy. thin cross-bdd. & cross-lam., vy. calc., f.g. SS w abundant plant frags., brachiopod casts & pelecypod casts.
	Datum for stratigraphic cross-section
13 m	gy., vy. thin cross-bdd. & cross-lam., f.g. SS w abundant f. to c. plant frags. & burrows and occ. oscill. ripple marks.
0.6 m	olive gy., thin bdd., calc., f.g. SS w sh. partings and interference ripple marks.
3 m	greenish tan, mottled, slty. MUDSTONE.
11 m	gy., vy. thin, shallow cross-bdd., f.g. SS w abundant plant frags., grading downward into olive gray, f.g. SS.
5 m	gy., thin & vy. thin, shallow cross-bdd., sltly. calc. SLTST, in part w small pelecypods, vy. f. burrows & plant frags. and occ. calc. nodules.

Base of Section

Section 6

Table 9-11

Top of Section

Thickness	Description
11 m	gy.-gn., cross-bedded, f. - m.g. SS
14 m	rd. SLTST & MDST
5 m	gy.-gn., cross-bedded, fining-upwards SS
17 m	rd. SLTST. & MDST. w root stigmaria and burrow tubes
8 m	gy.-gn., cross-bedded, f.-m.g. SS w CGLT lenses
8 m	rd. SLTST w grn. mottle patterns, root stigmaria at top
9 m	gy., cross-bdd; cgltc. f. - m.g. SS; plant fragments abundant
28 m	rd., vy. thin bdd., SLTST, in part w stigmaria & burrow tubes interbedded w thin beds of gy. cross-bedded v.f.g. SS
2 m	lt. tan, "clean," vy. thin & thin cross-bdd. f.g. SS
8 m	gy., vy. thin cross-bdd., f.g. SS
5 m	gy., vy. thin cross-bdd., f.g. SS
57 m	rd. and grn., in part w green mottling, in part slightly calc. and with root stigmaria, SLTST and MDST
2 m	gy. & gn.-gy., cross-bdd., f.g. SS w abundant plant frags. and a few "coaly" seams
10 m	rd. SLTST and MDST

Base of Section

NOTE: Plant fragments ubiquitous in this section.
No marine fossils observed.

Use base of 2 m "clean" sandstone as datum for your stratigraphic cross-section.

Section 7

Table 9-12

Procedure I: Plotting Stratigraphic Cross-Section

1. The cross-section should be drawn at a horizontal scale of 1 cm = 5 km and a vertical scale of 1 cm = 10 m

2. Assume that the cross-section trends East-West for a distance of 138 km.

3. The spacing between adjacent sections (running from west to east) is:

1 - 2 = 11 km	2 - 3 = 28 km	3 - 4 = 25 km
4 - 5 = 28 km	5 - 6 = 22 km	6 - 7 = 23 km

4. Plot the lithologic units for each section in relief style. (See Fig. 9-1 for the appropriate symbols.) The column should not exceed 1.5 cm in width.

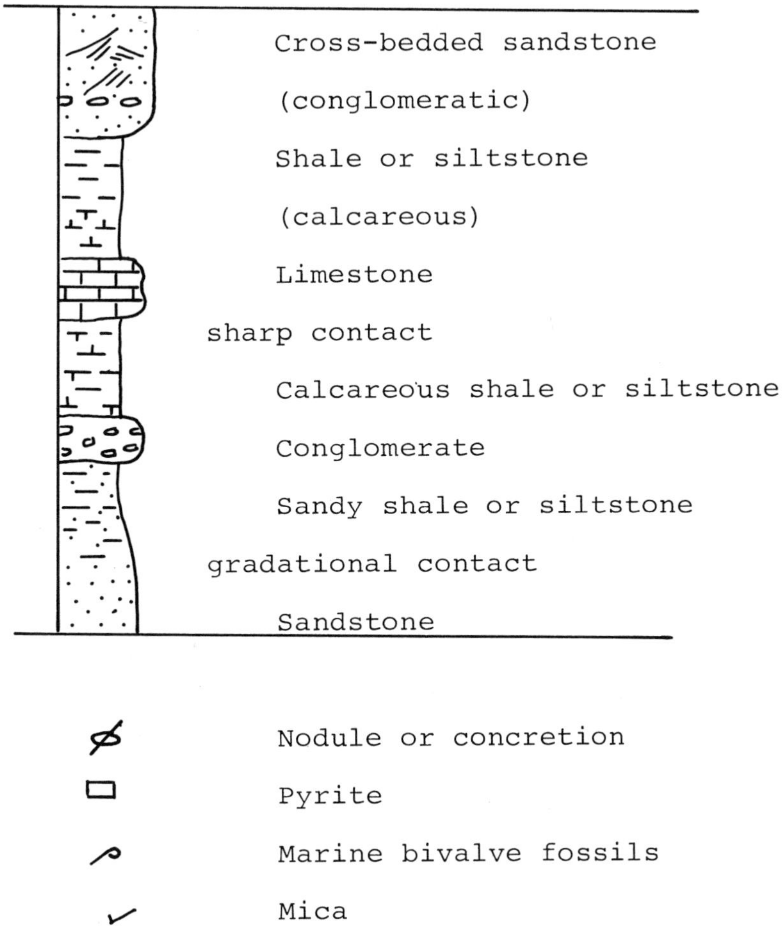

Standard Stratigraphic Symbols

Figure 9-1

Procedure II: Designation of Facies

1. Study the relationships of the lithologic units that appear on your stratigraphic cross-section.

2. Using the characteristics of the alluvial facies, tidal facies, nearshore facies and offshore facies outlined in Tables 9-2 through 9-5, establish which of these are represented on your cross-section.

3. After you feel that you have worked out the spatial relationships of the sedimentary facies, draw contact lines which separate them. A color pattern for each will help to bring out the configuration of the facies.

Procedure III: Interpretation of Facies

1. Using your stratigraphic cross-section, interpret the geological history of the area in terms of (a) tectonics of the source area, (b) tectonics of the depositional basin, and (c) transgression versus regression.

2. Summarize your interpretation in the space below.

Darned if I know. It
must have been a good
hunch.

E. DeGolyer, when asked
how he located Potrero
del Llano #4, which pro-
duced 130 million barrels
of oil.

EXERCISE 10
OIL AND GAS EXPLORATION PROJECT

This exercise is designed to simulate the geological and economic conditions usually encountered in oil exploration. The oil and gas fields in the area covered by the base map are largely structurally controlled, but not all highs are productive. Although one of the principal goals of the exercise is to drill productive wells and make a profit, some effort should be made to explore and exploit more than one structure, if possible, so as to gain geologic perspective. The ability to contour structural data is vital and a controlled imagination is frequently rewarded.

Exercise 10 has been constructed so that it will continue for more than a single laboratory period. Your instructor will provide information on the time framework that is appropriate for your class.

Procedure

Each operator (student) is to work completely alone, treating his geological information and structural interpretation as strictly confidential. Your laboratory instructor will serve as the source of all geological data, and will supply oil and gas production information. Each operator should keep track of his own financial status and plot all acquired and interpretational information on his base map. Your base map is in the pocket following page 196.

1. At the beginning, 27 points on the structure datum horizon are supplied free. These are to be located and contoured.
2. Each operator begins with $400,000, and stays in the exercise unless he becomes bankrupt. If he exhausts his original capital he may auction his holdings, or enter into a deal with another operator providing all operators agree to the terms. He may then continue.
3. New structural data is obtained by requesting seismic line at $10,000 per mile with a five mile limit per day, and by requesting information for up to 20 individual points at $1,000 each. This is a daily allotment and expires daily. Submit requests to your instructor on properly indexed overlay material. Seismic line data is supplied to the operator as datum points where the line intersects contours. The number of points actually received depends upon the orientation of seismic line with respect to the structural trend. Seismic line may be taken only

along the edges of sections. Individual points may be requested anywhere, provided they are taken at the center of a 40 acre tract. The map should be contoured at a 50-foot interval. All points are in feet below sea level.

4. At a set time each day, bids are submitted on desired acreage on a "dollars per acre" basis. Identify acreage for the instructor by citing town, range, section, and subsection divisions on overlay material. Bids will be immediately compared and winners and prices compiled on a property map which is public information. All operators should keep their maps up-to-date with this information.
5. The following day, and any time thereafter, wells may be drilled at a cost of $40,000 each. Structure data of wells drilled is supplied with-without additional cost. Oil producers yield $75,000 per day. Gas producers yield $30,000 per day. Production information is released to all operators each day after land bids are awarded and posted. Your instructor will post production on the land map.
6. A 20-acre spacing (center of a 40-acre tract) is minimum on most of the fields. In tight structure fields 10-acre spacing (4 wells on a 40-acre tract) is allowed, but such fact will be indicated only to landholding operators in these fields after they have brought in one producing well.
7. Until one producing well has been located within one mile of a section which is being bid upon, only that property shown by shading on this diagram will be available for purchase. After location of a well within one mile, all acreage is available.

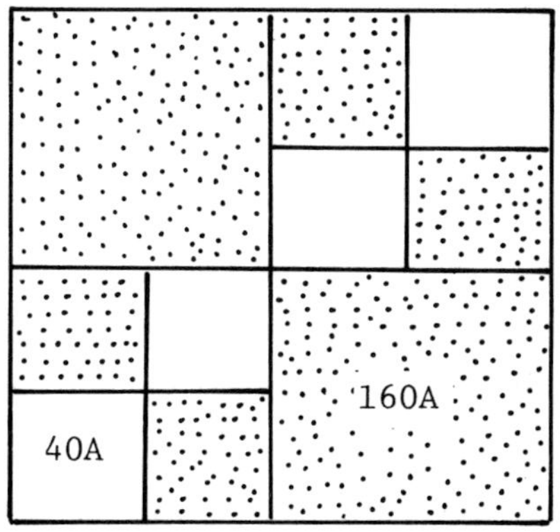

8. No more than 1600 acres may be bid upon per day.
9. The following points on the structure datum horizon are provided free of charge. Plot these data on your base map.

T15NR5E	S8	NENE	-3160'
	S10	NESE	-3070
	S12	NESW	-2880
	S20	SWSW	-3385
	S22	SESE	-3200
	S33	SWSW	-3600
	S35	NESE	-3405

T15NR6E	S4	SWSW	-3130
	S12	NWSW	-2960
	S16	NWNW	-3110
	S20	SWSE	-3180
	S26	NWNW	-3100
	S31	SWSE	-3215
	S34	SWSW	-3150
	S36	SESE	-2980
T14NR5E	S5	SWSW	-3620
	S11	NESW	-3290
	S17	SESE	-3760
	S14	SWSW	-3475
	S29	SWSE	-3700
	S35	NWNW	-3495
T14NR6E	S8	NWSW	-3255
	S16	NENE	-3090
	S19	NWSE	-3340
	S24	NWSW	-3200
	S34	NWNW	-3225
	S31	SWNE	-3360

GOOD LUCK!

REFERENCES CITED

(but not listed within body of individual exercises)

Friedman, G. M.; Amiel, A. J.; Braun, Moshe; and Miller, D. C., 1973, Generation of carbonate particles and laminites in algal mats - example from sea-marginal hypersaline pool, Gulf of Aqaba, Red Sea: Am. Assoc. Petroleum Geologists, Bull., v. 57, p. 541-557.

Goldberg, M.; and Friedman, G. M., 1974, Paleoenvironments and paleogeographic evolution of the Jurassic System in Southern Israel: Geol. Survey Israel, Bull. 61, 44 p.

Grabau, A., 1913, Principles of stratigraphy: New York, A. G. Seiler & Co., 1185 p.

Hubert, J. F., 1971, Analysis of heavy-mineral assemblages, Ch. 19, p. 453-478, in Carver, R. E., ed., Procedures in sedimentary petrology: New York, Wiley-Interscience, 653 p.

McKinney, T. F.; and Friedman, G. M., 1970, Continental shelf (sic) sediments of (sic) Long Island, New York: Jour. Sedimentary Petrology, v. 40, p. 213-248.

Okada, H., 1971, Classification of sandstone: analysis and proposal: Jour. Geology, v. 79, p. 509-525.

Triplehorn, D. M., 1966, Morphology, internal structure and origin of glauconite pellets: Sedimentology, v. 6, p. 247-266.

Visher, G. S., 1965, Fluvial processes as interpreted from ancient (sic) and recent fluvial deposits, p. 116-132, in Middleton, G. V., ed., Primary sedimentary structures and their hydrodynamic interpretation: Tulsa, Okla., Soc. Econ. Paleontologists and Mineralogists, Spec. Pub. No. 12, 265 p.

__________, 1969, Grain size (sic) distributions and depositional process: Jour. Sedimentary Petrology, v. 39, p. 1074-1106.

GLOSSARY

Accessory mineral. A mineral that occurs in relatively small quantities in a rock and whose presence or absence does not affect its analysis.

Allochem. A general term for any kind of sand-size or larger particle, in limestones, that was derived from within a basin of deposition.

Anhedral. Texture of crystals having boundaries that are not crystal faces.

Anthracite. A hard, black, lustrous coal containing a high percentage of fixed carbon and a low percentage of volatile matter.

Arenite (Grabau classification). A sedimentary rock consisting of particles of sand size.

Argillaceous. Applied to all rocks or substances composed of clay, or having a notable proportion of clay in their composition, as roofing slate, shale, etc.

Arkose (Krynine definition). A sandstone containing more than 30 percent feldspar.

Authigenic. A mineral that has grown in place subsequent to the formation of the sediment or rock of which it constitutes a part.

Beachrock. A friable to well-cemented sedimentary rock, formed in the intertidal zone in a tropical or subtropical region, consisting of calcareous debris (detrital and/or skeletal) cemented with calcium carbonate.

Bituminous coal. Coal that ranks between subbituminous coal and semibituminous coal and that contains 15-20% volatile matter.

Bog-iron deposits (ore). A general term for a soft, spongy, and porous deposit of impure hydrous iron oxides formed in bogs, marshes, swamps, peat mosses, and shallow lakes by precipitation from iron-bearing waters and by the oxidizing action of algae, iron bacteria, or the atmosphere.

Caliche. A limestone, precipitated as surface or nearsurface crusts and nodules by the evaporation of soil moisture in arid and semiarid climates. Syn.: nari, calcrete.

Chamosite. A greenish-gray or black mineral of the chlorite group: $(Mg, Fe)_3Fe_3^{+3}\,(AlSi_3)O_{10}(OH)_8$. It is the monoclinic dimorph of orthochamosite, and is an important constituent of many oölitic and other bedded iron ores.

Chemical weathering (decomposition). Process by which chemical reactions change rocks and minerals into new chemical combinations that will remain indefinitely without further change under conditions at or near the surface of the Earth.

Clarain. A coal having a smooth surface when broken at right angles to the bedding plane; such broken faces have a pronounced gloss or shine and silky luster.

Coalification. The increase in the relative content of carbon and calorific value and decrease in contents of moisture, volatile matter, and oxygen experienced by carbonaceous material in being transformed from peat to lignite to bituminous coal to anthracite.

Coccolithophore. Any of numerous, minute, mostly marine, planktonic biflagellate protists having brown pigment-bearing cells that at some phase of their life cycles are encased in a sheath of coccoliths to form a complex calcareous shell.

Coprolite. An animal's fecal particle, commonly 1 to 15 cm in length, light to dark brown or black in color, and usually elongate to ovoid in shape.

Crinoid. Any pelmatozoan echinoderm belonging to the class Crinoidea, characterized by quinqueradiate symmetry, by a disk-shaped or globular body enclosed by calcareous plates and from which appendages, commonly branched, extend radially, and by the presence of a stem, or column, more common in fossil than in living forms.

Cryptocrystalline. Said of the texture of a rock consisting of or having crystals that are too small to be recognized and separately distinguished even under the ordinary microscope.

Cumulative-frequency curve. A frequency curve constructed by starting at one end of a distribution and by proceeding toward the other end of the distribution by adding progressively the sums of the percentages of each succeeding class.

Decantation. The process of pouring off a liquid gently without stirring up the contained sediment.

Depositional environment. A natural geographic entity in which sediments accumulate.

Detrital. Pertaining to or formed from detritus, a collective term for loose rock and mineral material that is worn off or removed directly by mechanical means, as by disintegration or abrasion.

Diamictite. A nonsorted, noncalcareous, terrigenous sedimentary rock composed of sand-size and/or larger particles dispersed through a fine-grained matrix. Compare diamicton.

Diamicton. A nonsorted, noncalcareous, terrigenous sediment composed of sand-size and/or larger particles dispersed through a fine-grained matrix. See also diamictite.

Dry-sieving. The process of sieving a sediment sample without using water.

Durain. A dull coal lacking luster and having an earthy appearance.

Epigenetic dolostone. A dolostone that has formed by localized replacement of limestones along postdepositional structural features, such as faults and fractures.

Euhedral. Texture of a crystal wholly bounded by crystal faces.

Evaporites. Sedimentary deposits composed of minerals precipitated from brines concentrated by evaporation. As applied to sedimentary rocks, evaporites designate a group of authigenic rocks whose minerals were precipitated from brines concentrated by evaporation.

Extrabasinal rocks. Rocks composed of sedimentary particles derived from weathering and erosion of a land mass.

Fabric. The arrangement of particles or crystals in a sedimentary deposit in terms of orientation by form, mineral composition, packing, and sorting.

Facies. The lithologic and biologic characteristics, of a sedimentary deposit, imparted by the depositional environment.

Fines. Very small particles, especially those smaller than the average in a mixture of particles of various sizes.

First cycle sediments. Sediments consisting of particles derived directly from the weathering and erosion of an igneous or metamorphic land area.

Fissility. The property of splitting easily into thin layers parallel to the bedding.

Frequency (distribution) curve. A curve that graphically represents a frequency distribution.

Fusain. Carbonized wood resembling charcoal.

Fusulinid. Any foraminifer belonging to the suborder Fusulina, characterized by a calcareous microgranular test.

Glauconite. A dull-green, amorphous, and earthy or granular mineral of the mica group, commonly regarded as the iron-rich analogue of illite. It is an indicator of very slow sedimentation.

Goethite. A yellowish, reddish, or brownish-black mineral: α-FeO(OH). Goethite is the commonest constituent of many forms of natural rust or of limonite, and it occurs especially as a weathering product in the gossans of sulfide-bearing ore deposits.

Graywacke (Krynine definition). A sandstone composed of angular quartz (and chert) particles and abundant metamorphic rock fragments, with little or no cement and feldspar, and containing more than 12 to 17 percent micas and chlorite (either in the clay matrix or as metamorphic rock fragments).

Greenalite. An earthy- or pale-green mineral: $(Fe^{+2}, Fe^{+3})5\text{-}6\text{-}Si_4O_{10}(OH)_8$. It occurs in small ellipsoidal granules in cherty rock associated with the iron ores of the Mesabi district, Minn. Greenalite resembles glauconite in appearance, but contains no potassium.

Heavy. See Heavy mineral.

Heavy mineral. A detrital mineral in a sedimentary rock, having a specific gravity higher than a standard (usually 2.85), and commonly occurring as a minor constituent or accessory mineral of the rock ($<$ 1% in most sands).

Histogram. A vertical bar-graph representation of a frequency distribution in which the height of bars is proportional to frequency of occurrence within each class interval.

Hydrozoan. Any coelenterate belonging to the class Hydrozoa, characterized by forms that are both polypoid and medusoid or by exclusively medusoid forms and by the absence of nematocysts and a stomodaeum.

Hypidiotopic. A fabric in sedimentary rocks in which most of the constituent crystals are subhedral.

Intrabasinal rocks. Sedimentary rocks whose constituent particles were derived from within the basin of deposition.

Idiotopic. A fabric, in sedimentary rocks, in which the predominant constituents are euhedral crystals.

Ignimbrite. A nonsorted pyroclastic rock deposited from a nuee ardente (q.v.) while the particles were still in a plastic condition.

Intraclast. A sand-size or larger nonskeletal particle, texturally analogous to a rock fragment (q.v.), broken from consolidated or hardened carbonate materials that are accumulating within the basin of deposition. See peloid; compare grapestone.

In situ. In place.

Lights. Rock-forming minerals of a detrital sedimentary rock, having a specific gravity lower than a standard (usually 2.85).

Lignite. A brownish-black coal that is intermediate in coalification between peat and subbituminous coal.

Lime mud. The unconsolidated micritic component of a carbonate sedimentary deposit.

Lithification. The conversion of loose particles to masses of solid rock.

Lussatite. A fibrous form of cristobalite resembling chalcedony.

Lutecine. Fibrous chalcedony characterized by inclined extinction and by fibers that are seemingly elongated about 30° to the c-axis.

Lutite (Grabau classification). A sedimentary rock consisting of particles of clay and silt sizes.

Mean. The midpoint of a distribution, the point that divides the area under the frequency distribution curve into two equal parts.

Mechanical weathering (disintegration). Process by which physical forces break down or reduce rock to smaller and smaller fragments, and involving no chemical change.

Megaquartz. In chert, quartz consisting of crystals larger than 35 μ in size.

Micrite. Lithified, mechanically deposited lime mud, which occurs either as a matrix among sand-size particles in a limestone or as the only particle in a fine-grained limestone. In micrite, the particle sizes are <4 microns, usually 1 to 3 microns.

Micron (μ). One millionth of a meter, one thousandth of a millimeter.

Microquartz. In chert, quartz consisting of crystals smaller than 35 μ in size.

Microspar. A mosaic of neomorphic crystals having diameters ranging from 4 to 10 microns or even to 50 microns.

Mode. The most-abundant class in population; thus, the highest point on a frequency-distribution curve.

Moment. The average or sum of the deviations or some power of the deviations of the elements of a frequency distribution from a specified norm.

Montmorillonite. A group of expanding-lattice clay minerals generally derived from alteration of ferromagnesium minerals, calcic feldspars and volcanic glasses.

Mud. A collective term for sediments whose particle sizes lie within the range of various classes of silt and clay.

Mutlipopulational. In a particle-size frequency distribution, having several modes. This suggests that the particles were derived from several parent deposits.

Neomorphism. A transformation between one mineral and itself or a polymorph resulting in the growth of new crystals that are larger or simply different in shape from the previous particles or crystals.

Oligomictic conglomerate. A conglomerate whose particles consist of only a single kind of rock.

Oöid. A spherical or elliptical particle, smaller than 2 mm in diameter, typically composed of carcium carbonate and having a central nucleus surrounded by a rim consisting of more than one layer, that displays concentric or radial fabric. Compare pisolite.

Opaque. A mineral particle which has the characteristic of not letting light pass through it.

"Orthochems" (Folk classification). The material that binds together the allochems of limestones, and that includes cement (sparry calcite) and micrite. Syn.: orthochemical constituents.

Packing. The manner of arrangement or spacing of the solid particles in a sediment or sedimentary rock.

Pellet. A sand-size, structurally homogeneous, spherical or ellipsoidal nonskeletal particle, consisting of the nondigested residue, commonly calcium-carbonate mud, secreted anally by deposit-feeding organisms. See also peloid and intraclast.

Peloid. A sand-size nonskeletal particle resembling a pellet but for which no particular origin is implied. See pellet and intraclast.

Percentile. Any of the values in a series dividing the distribution of the individuals in the series into one hundred groups of equal frequency.

Phi scale. A geometric (octave) grade scale of the sizes of sediment particles using class boundaries based on negative logarithms to base 2. The phi scale relates to a millimeter scale as follows:

$$\phi = -\log_2 \text{size (mm)}$$

Pisolite. A spherical or elliptical nonskeletal particle, containing two or more concentric layers of carbonate, and exceeding 2 mm in diameter. Compare oöid.

Poikilotope. A large crystal, in a poikilotopic fabric, set among smaller crystals and enclosing original particles or newly formed crystals.

Polymictic conglomerate. A conglomerate whose particles include many kinds of rocks.

Polymodal distribution. A frequency distribution characterized by two or more localized modes, each having a higher frequency of occurrence than other immediately adjacent individuals or classes.

Poly-varietal. Having more than one variety of constituent particle composition. For example, a poly-varietal sand.

Porphyrotope. A large crystal, in a porphyrotopic fabric, enclosed among smaller particles or crystals.

Provenance. The scientific study of the origins of sedimentary materials; usually refers to determinations of the locations and compositions of the parent deposits from which sedimentary particles were eroded.

Pseudospar. A mosaic of neomorphic crystals having diameters larger than those of microspar ($>$ 10 to 50 microns).

Pyroclastic rock. Rock formed from tephra.

Pyroclastic particles. Clastic sedimentary particles formed by volcanic explosion or aerial expulsion from a volcanic vent.

Quartile deviation. The 25th and 75th percentiles are quartile deviations. These represent the central part of the size frequency-distribution curve and hence involve the bulk of the particles.

β quartz (Beta quartz). The polymorph of quartz that is stable from 573°C to 870°C. It occurs as phenocrysts in quartz porphyries, graphic granite and granite pegmatites.

Quartzine. Fibrous chalcedony characterized by fibers having a positive crystallographic elongation (fibers parallel to the c-axis).

Radiolarian. Any actinopod belonging to the subclass Radiolaria, characterized mainly by a siliceous skeleton and a marine pelagic environment. Their stratigraphic range is Cambrian to present. In some classifications the radiolarians are grouped with the rhizopods.

Rock fragment. A sedimentary particle having recognizable characteristics of its parent deposit.

Roundness. The attribute of form of a particle that is related to the sharpness or curvature of edges and of corners.

Rudite (Grabau classification). A sedimentary rock consisting predominantly of particles of pebble, cobble, and/or bounder sizes. A conglomerate or sedimentary breccia of modern usage.

Saltation. The bouncing motion of particles being moved by a fluid.

Second cycle sedimentary deposit. A sedimentary deposit consisting of particles that have undergone two stages of erosion, transportation and deposition.

Sedimentary (depositional) environment. A natural geographic entity in which sediments accumulate.

Settling rate. The velocity at which suspended particles are deposited.

Siderite. A rhombohedral mineral, $FeCO_3$. It is isomorphous with magnesite and rhodochrosite, and commonly contains magnesium and manganese. It is a valuable ore of iron.

Siltite. A sedimentary rock consisting of silt-size particles.

Skeletal debris. Material derived from organisms, consisting of the hard parts secreted by the organisms.

Skewness. Deviation of a given distribution from the symmetry of a normal distribution.

Sorting. The selection, during transport, of particles according to their sizes, specific gravities, and shapes. Statistically, sorting is expressed as the dispersion of the particle sizes about the mean (q.v.).

Source terrane. The land area from which sedimentary particles were eroded; the nature of source terrane can often be inferred from the composition of a sedimentary deposit.

Sparry calcite. A cement consisting of a mosaic of calcite crystals larger than those of micrite.

Sphericity. The attribute of form that is related to the degree to which the three mutually perpendicular axes of a particle, which represent its longest (L) intermediate (I), and shortest (S) dimensions, approach equality with one another. In theory, sphericity can be expressed numerically by the ratio between the surface area of a particle and the surface area of a sphere having a volume equal to that of the particle. A simpler approximation, the maximum projection sphericity is computed by

$$\left(\frac{S^2}{LI}\right)^{1/3}$$

Sphericity is geometrically independent of roundness (q.v.).

Standard deviation. A departure from the middle of a normal distribution equal to 34 percent of the total population. The symbol for standard deviation is σ. The numerical value of the standard deviation for any population is computed by taking the square root of the second moment about the mean. Stated mathematically, this is

$$\sigma = \left[\frac{\sum f(m_\phi - \bar{x})^2}{100}\right]^{1/2}$$

where $\bar{x}$ = phi value of the mean, and m_ϕ = phi value of the midpoint of the distribution class.

Stokes' law. A formula that expresses the rates of settling of spherical particles in a fluid: $V = Cr^2$, where V is velocity (Cm/Sec), r is particle radius (cm), and C is a constant which takes into account gravity, fluid and particle densities, and fluid viscosity.

Stromatolite. A laminated lithified deposit consisting of alternating layers of material formed by microorganisms, usually algae, and of extraneous particles that have been cemented. The material formed by microorganisms may consist of soft organic matter, or its fossil analogue, or of thin solid layers of precipitated carbonate material. The extraneous particles can consist of anything, but among these fine-grained carbonate particles typically predominate.

Subhedral. Texture of crystals partly bounded by crystal faces.

Sucrosic. Dolostones having "sugary" appearance. Commonly the fabric of such dolostones is idiotopic.

Surface texture. The micromorphologic characteristics of the external surfaces of sedimentary particles.

Tephra. A general term for all particles exploded from volcanoes.

Terrigenous particles. Clastic sedimentary particles derived from weathering and erosion of the continents.

Texture. The size, shape and arrangement of the constituent particles of a sedimentary rock.

Tillite. Till that has become rock.

Travertine. A hard, dense, finely crystalline, compact or massive but often concretionary limestone, of white, tan or cream color formed by rapid chemical precipitation of calcium carbonate from solution in surface and ground water.

Tufa. A chemical sedimentary rock composed of calcium carbonate, formed by evaporation as a thin, soft crust around the mouth of a hot or cold calcareous spring or seep, or along a stream bank or lake shore.

Vitrain (anthraxylon). A bright, glossy, vitreous or glassy looking coal having conchoidal fracture.

Wacke. A "dirty" sandstone that consists of a mixed variety of angular and unsorted or poorly sorted mineral and rock fragments and of an abundant matrix of clay and fine silt.

Wet-sieving. The process of sieving a sediment sample using water.

Xenotopic fabric. An arrangement of anhedral crystals in sedimentary rocks.